GROW YOUR MIND

Work hard
Get Dirty

We should not teach children the sciences;
but give them a taste for them.

—JEAN JACQUES

THE GARDEN STEM

Science. Technology. Engineering. Math

CURRICULUM FOR YOUR GARDEN CLASSROOM

Grades K-8

Address: The College of Charleston, 66 George Street, Charleston, SC 29424
Email: Thompsonom@cofc.edu

Order copies at online retailers

Print ISBN : 978-1-949979-97-8

Printed in the United States of America

Managing Editor: Olivia M. Thompson, Ph.D., M.P.H.
Lead Editorial Assistant: Greta J. Cappelmann, M.P.H.c.
Editorial Assistant: Brittney Harrington, M.Ed.
Book Designer: Torborg Davern Design
Photographer: Leslie McKellar
Technical Writer: Stephanie Alexander, M.A.
Curriculum Specialist: Katie Donohoe, M.S.

Funding for this project provided by The Boeing Company

First Edition

TEXTBOOK AUTHORS:

KINDERGARTEN—8TH GRADE CURRICULUM

Morgan Walko-Bell, *Teacher*
Jerry Zucker Middle School of Science,
North Charleston, South Carolina

Katie Donohoe, *Teacher*
Founder, Growing Minds,
Mount Pleasant, South Carolina

Rose M. Fashion, M.Ed., *Teacher*
Berkeley Elementary School,
Moncks Corner, South Carolina

Lauren P. Fields, *Teacher*
Murray-LaSaine Elementary School,
Charleston, South Carolina

Kaitlin Fitzgerald, *Program Manager*
The Green Heart Project,
Charleston, South Carolina

Samantha Gentrup, M.Ed., *Teacher*
West Ashley Middle School,
Charleston, South Carolina

Erika L. Carasik-Gilly, M.S., *Teacher*
Morningside and Jerry Zucker Middle School,
North Charleston, South Carolina

Karen M. Latsbaugh, *Teacher*
East Cooper Montessori Charter School,
Mount Pleasant, South Carolina

Shaney B. Odgers, M.Ed., *Teacher*
St. Stephen Elementary School,
St. Stephen, South Carolina

Leila O Neal, M.Ed., *Teacher*
Whitesville Elementary School,
Moncks Corner, South Carolina

Karrilynn Savage, *Teacher*
St. Stephen Middle School,
St. Stephen, South Carolina

Katherine E. Schmenk, M.A., *Teacher*
Haut Gap Middle School,
Johns Island, South Carolina

Kristen Shawen France, M.Ed., *Teacher*
Alston Middle School,
Summerville, South Carolina

Ashley H. Steadly, M.A., *Teacher*
Ashley River Creative Arts Elementary School,
Charleston, South Carolina

Crystal M. Verch, M.Ed., *Teacher*
St. Stephen Elementary School,
St. Stephen, South Carolina

Melissa Nehez-Wingard, *Teacher*
Mitchell Math and Science Elementary School,
Charleston, South Carolina

David Wingard, *Teacher*
Mitchell Math and Science Elementary School,
Charleston, South Carolina

HARVEST TASTING GUIDE

Dana Mitchel, R.D., L.D.,
Nutrition Health Educator
Lowcountry Food Bank,
North Charleston, South Carolina

CURRICULUM WRITING WORKSHOP PLANNING COMMITTEE

Amy L. Dabbs, M.S., *Urban Horticultural Agent*
Clemson University Extension,
Charleston, South Carolina

Maggie Dangerfield, M.S., M.P.A., *Sustainability Coordinator*
Charleston County School District,
Charleston, South Carolina

Drew J. Harrison, *Director*
The Green Heart Project,
Charleston, South Carolina

Jennifer Schlette (Planning Committee Lead), *4-H Agent*
Clemson University Extension,
Charleston, South Carolina

Olivia M. Thompson, Ph.D., M.P.H., *Assistant Professor and Director*
Division of Public Health Sciences,
The Mayor Joseph P. Riley Jr. Center for Livable Communities
College of Charleston,
Charleston, South Carolina

TABLE OF CONTENTS

LESSONS FOR FOURTH GRADE

LESSONS FOR FIFTH GRADE

LESSONS FOR SIXTH GRADE

LESSONS FOR SEVENTH GRADE

LESSONS FOR EIGTH GRADE

HARVEST TASTING GUIDE

SOUTH CAROLINA SEASONAL RAINBOW 561

WELCOME TO THE GARDEN STEM

Dear South Carolina Educator:

Interest and success in science and science-related disciplines can often be traced back to students' experiences during elementary- and middle-school years. Students' natural curiosity and enthusiasm for learning science as well as any other subject can be further cultivated through experiential and inquiry-based instruction. The National Science Teachers Association describes how these learning experiences allow children to develop skills, such as problem solving, that will help them succeed in "an increasingly scientific and technological world" (National Science Teachers Association, 2002). *The Garden STEM* is a science-intensive, standards-based curriculum that integrates garden-based learning experiences into lessons designed to teach K-8 students in S.T.E.M. science, technology, engineering, and mathematics using the Five E Instructional Model, which is briefly described in the Lesson Structure section below.

The Garden STEM was designed to help South Carolina students excel academically while at the same time improving their psychosocial development.

We hope that you enjoy this resource!

Sincerely,
The Garden STEM Team

GETTING STARTED

GARDEN REQUIREMENTS

This garden-based curriculum may be integrated into any type of gardening program – whether your school garden is a container garden, indoor garden, raised-bed, or traditional garden/farm, *The Garden STEM* can bridge classroom learning to the garden. Creating positive learning experiences is key to helping students learn to love gardening and nature experiences. Barriers to accessibility such as poorly designed gardens or unsafe conditions can be avoided by connecting to available resources, so please contact us and we will help you find the needed technical assistance.

TIME REQUIREMENTS / LESSON DURATION

Each lesson is designed to be taught over the course of approximately 1 session or class period, unless otherwise stated.

EDUCATIONAL STANDARDS

The lessons, suitable for kindergarten through eighth grade, are aligned to South Carolina academic standards and performance indicators in science, technology, engineering, visual and performing arts, and health education. Additionally, lessons are aligned with the new South Carolina College and Career-Ready Standards for Mathematics and English Language Arts.

LESSON STRUCTURE

The Garden STEM lesson plans follow the 5E Instructional Model, a popular teaching and learning sequence that has proven to effectively develop students' understanding of subject matter. This model works by dividing each lesson into 5 clear stages, providing transparency and clear instruction for both students and teachers. The Instructional Model is separated thus:

Engagement: This section allows teachers to gauge prior understanding of the subject, and frame the lesson s focus.

Exploration: This section gets students involved in the topic, and provides students with activities to test their knowledge and skills.

Explanation: This section serves to provide students with an opportunity to communicate and discuss new understandings.

Elaboration: This section allows students to develop deeper and broader understandings and/or skills, and apply those to new situations.

Evaluation: This section provides both students and teachers a chance to assess how much learning has taken place.

This method has become very popular with science curricula in particular, and has shown to improve students' mastery of the subject matter and enhance scientific reasoning. This method also has been shown to foster teamwork abilities and practical skills that students will use for the rest of their lives.

For additional information please see:

Bybee, Rodger W. et. al., "The BSCS 5E Instructional Model: Origins, Effectiveness, and Applications", July 2006, Colorado Springs, CO, www.bscs.org.

Verstynen, Shannon, "5Es Overview: The 5E Instructional Model", March 2013, www.nasa.gov.

Barufaldi, Jim, "The 5E Model of Instruction", July 2002, Austin, TX.
http://www.wisd.org/users/0001/docs/GVC/5E 20Model.pdf.

LESSONS FOR KINDERGARTEN

KINDERGARTEN LESSON PLAN
Working Worms

SUBJECT / GRADE LEVEL: Kindergarten - Exploring Organisms and the Environment.

ACADEMIC STANDARDS AND PERFORMANCE INDICATORS:

ELA:

RI.MC.5.1 With guidance and support, ask and answer who, what, when, where, why, and how questions about a text; refer to key details to make inferences and draw conclusions in texts heard or read.

RI.LCS.9.1 With guidance and support, ask and answer questions about known and unknown words.

RI.LCS.8.2 With guidance and support, use front cover, title page, illustrations / photographs, fonts, glossary, and table of contents to locate and describe key facts or information; describe the relationship between these features and the text.

RL.LCS.10.1 With guidance and support, ask and answer questions about known and unknown words.

RL.RC.13.1 Engage in whole and small group reading with purpose and understanding.

RL.MC.5.2 With guidance and support, ask and answer questions to make predictions using prior knowledge, pictures, illustrations, title, and information about author and illustrator.

W.MCC.2.1 Use a combination of drawing, dictating, and writing to compose informative/explanatory texts that name and supply information about the topic.

W.MCC.2.2 With guidance and support, plan, revise, and edit building on personal ideas and the ideas of others to strengthen writing.

C.MC.1.4 Participate in conversations with varied partners about focused grade level topics and texts in small and large groups.

C.MC.2.1 With guidance and support, recall information from experiences or gather information from sources to ask and answer questions.

C.MC.2.2 With guidance and support, participate in shared research exploring a variety of texts; express opinions and talk about findings.

C.MC. 3.2 Use appropriate props, images, or illustrations to support verbal communication.

Math:

K.NS.1 Count forward by ones and tens to 100.

K.NS.7 Determine whether the number of up to ten objects in one group is more than, less than, or equal to the number of up to ten objects in another group using matching and counting strategies.

LESSON OBJECTIVES:

TSW understand that decomposition is an important process that takes place in the garden. TSW identify the decomposers that live in their garden.

SOFT SKILLS: Emotional intelligence, work ethic, teamwork, positive attitude.

****Differentiation strategies such as flexible grouping teaming and/or pairing appropriately can be used to meet diverse learning needs. ESOL students can work with the teacher. Students with physical needs should be assisted by the teacher, assistant, shadow, or another student.*

MATERIALS:

- Trowel or shovel
- Paper plates
- Paper cups
- Pens/paper
- Magnifying glass
- Large jar
- *Diary of a Worm* by Doreen Cronin
- *An Earthworm's Life* by John Himmelman
- *The Big Fat Worm* by Nancy Laan
- *Working Worms* student investigation guide (attached)
- *Working Worms* teacher guide (attached)

ENGAGEMENT:

- Teacher will pose the question, "**What does a worm do at work every day?**"
- Teacher will follow the engagement question by reading a worm book to the class. Suggestions include: *The Diary of a Worm* by Doreen Cronin, *An Earthworm's Life* by John Himmelman, or *The Big Fat Worm* by Nancy Laan.

EXPLORATION:

- Teacher will guide students through an exercise in which they will collect soil from the garden and examine decomposers.

EXPLANATION:

- Teacher will explain the vocabulary while students investigate the role of decomposers in the garden through inquiry based questioning.

ELABORATION:

- Students will practice their measurement skills while completing the measurement portion of the *Working Worms* student investigation guide.

EVALUATION:

- Students will conclude the *Working Worms* student investigation guide by drawing pictures of the decomposers they found.
- Teacher will observe reactions and interactions with other students and the decomposers.

RESOURCES

1. *Diary of a Worm* by Doreen Cronin.
2. *An Earthworm's Life* by John Himmelman.
3. *The Big Fat Worm* by Nancy Laan.
4. *Working Worms* handout (attached).

TEACHER GUIDE
WORKING WORMS

BACKGROUND:

Students should understand that decomposition is an important process that takes place in the garden. Students should identify the decomposers that live in their garden.

ENGAGEMENT:

- Teacher will pose the question, "**What does a worm do at work every day?**"
- Teacher will follow that engagement question by reading a worm book to the class. Suggestions include: *The Diary of a Worm* by Doreen Cronin, *An Earthworm's Life* by John Himmelman, or *The Big Fat Worm* by Nancy Laan.

EXPLORATION:

- Each student should dig up a soil sample and place it in a paper cup.
- Ask students to describe any living things that may have been in the soil when they scooped it up.
- **Are the living things helpful or harmful?**
- Each student should dump their soil sample on a paper plate.
- Ask students to use a magnifying glass to carefully examine the soil for decomposers.
- Describe to students the role of decomposers in the soil.
- Have each student place their decomposers in a jar on the teachers desk where they can be observed by the entire class. Cover the decomposers with a little bit of soil for protection.
- Students should collect more decomposers for examination. **This time were there different decomposers? Were there more or less of each one?**
- Keep exploring the soil in different parts of the garden and at varying depths until the class has a jar full of worms and other decomposers.

EXPLANATION:

- **Why are worms helpful to a garden?** Worms are a garden s best friend They eat dead and decaying plant waste and turn it into new soil for new plants.
- Worms are invertebrates, which means they don't have a backbone like humans.
- Worms do not have eyes, and they breathe through their skin.

- Vocabulary: *decay, plant waste, soil, creature, sample, fertile, decomposer, fewer/more, examine, habitat, underground, depth, worm, explore.*

ELABORATION:

- Students should practice measuring skills by completing the worm measurement portion of the *Working Worms* student investigation guide.
- Bury paper plates and cups in the garden. Estimate the length of time it will take for the items to compost with a little help from the decomposers. Check monthly until the objects are completely decomposed.
- As a class, create a diary for one of the creatures found in the garden.
- Be sure to return all the decomposers to the garden as well.

EVALUATION:

- Have students complete the *Working Worms* student investigation guide by drawing pictures of decomposers they found.
- Verbally share what each group or pair of students observed.
- Tell a story about the decomposers that the students found. Have students imagine what it is like to live underground. **What does it look like? How does it feel? What would you eat?**
- Encourage students to actively engage in the decomposer discussion.
- Teacher will observe reactions and interactions with other students and the decomposers.

WORKING WORMS

Name ____

Use the ruler to measure two different worms.

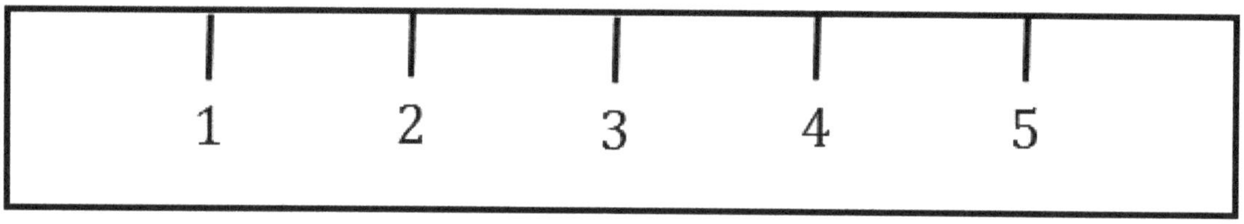

Worm 1 ☐   Worm 2 ☐

Which worm is the longest? (Circle the correct answer)

Worm 1   Worm 2

Draw a picture of two different decomposers.

KINDERGARTEN LESSON PLAN
Shake Up That Soil

SUBJECT / GRADE LEVEL: Kindergarten - Exploring Organisms and the Environment.

ACADEMIC STANDARDS AND PERFORMANCE INDICATORS:

ELA:

RI.MC.5.1 With guidance and support, ask and answer who, what, when, where, why, and how questions about a text; refer to key details to make inferences and draw conclusions in texts heard or read.

RI.LCS.9.1 With guidance and support, ask and answer questions about known and unknown words.

RI.LCS.8.2 With guidance and support, use front cover, title page, illustrations/ photographs, fonts, glossary, and table of contents to locate and describe key facts or information; describe the relationship between these features and the text.

RL.LCS.10.1 With guidance and support, ask and answer questions about known and unknown words.

RL.RC.13.1 Engage in whole and small group reading with purpose and understanding.

RL.MC.5.2 With guidance and support, ask and answer questions to make predictions using prior knowledge, pictures, illustrations, title, and information about author and illustrator.

W.MCC.2.1 Use a combination of drawing, dictating, and writing to compose informative/explanatory texts that name and supply information about the topic.

W.MCC.2.2 With guidance and support, plan, revise, and edit building on personal ideas and the ideas of others to strengthen writing.

C.MC.1.4 Participate in conversations with varied partners about focused grade level topics and texts in small and large groups.

C.MC.2.1 With guidance and support, recall information from experiences or gather information from sources to ask and answer questions.

C.MC.2.2 With guidance and support, participate in shared research exploring a variety of texts; express opinions and talk about findings.

C.MC. 3.2 Use appropriate props, images, or illustrations to support verbal communication.

Math:

K.MDA.3 Sort and classify data into 2 or 3 categories with data not to exceed 20 items in each category.

MATERIALS:

- Soil samples
- Hand trowels, small shovels, or large spoons
- Three large jars: spaghetti sauce jars work well
- Hand lens or magnifying glass
- Dishwashing detergent
- Sandwich bags
- *Soil* by Christen Ditchfield
- *Shake Up That Soil* student investigation guide (attached)
- *Shake Up That Soil* teacher guide (attached)

LESSON OBJECTIVES:

TSW explore the components of soil. TSW investigate the properties of soils including color, texture, smell, and size.

ENGAGEMENT:

- Teacher will pose the question, "**What is in the soil?**"
- Teacher will follow the engagement question by reading *Soil* by Christen Ditchfield to the class.

EXPLORATION:

- Teacher will guide students through an exercise in which they will collect soil from the garden and examine its color, texture, smell and size.

EXPLANATION:

- Teacher will explain the vocabulary while students investigate the properties of soil through inquiry based questioning.

ELABORATION:

- Students will practice their observation skills while completing the *Shake Up That Soil* student investigation guide.
- Teacher will conduct a classroom discussion in order for students to apply their knowledge of the components of soil.

EVALUATION:

- Students will turn in the *Shake Up That Soil* student investigation guide.
- Teacher will observe reactions and interactions with other students and the soil samples.

RESOURCES:

1. *Soil* by Christen Ditchfield.
2. *Shake Up That Soil!* handout (attached).

TEACHER GUIDE

SHAKE UP THAT SOIL

BACKGROUND:

Students should understand that soil is composed of living and non-living components. Students should identify the properties of soils including color, texture, smell, and size.

ENGAGEMENT:

- Teacher will pose the question, "**What is in the soil?**"
- Teacher will follow the engagement question with the reading of *Soil* by Christen Ditchfield.

EXPLORATION:

- Each student should use a sandwich bag to collect soil samples from three locations. These three locations should include the garden and non-garden area of school grounds and a third area of their choice.
- **DAY 1:** After collecting soil samples, students will explore the samples using hands, spoons, and lenses/magnifying glasses. Students can work individually, or in small groups.
- Ask students to describe the soil found in each of the three locations. Pose the questions below. Encourage students to use their senses to describe the soil, and have them address color, texture, and size.

  –How are the samples alike?

  –How are they different?

  –What is soil?

  –What is in soil?

  –What does soil feel like?

  –What does soil smell like?

  –Does all soil look the same?

  –How does soil help plants?

  –How does soil help animals?

  –What does the soil look like?

- Fill the three large jars about two-thirds full of water, then add soil samples from three distinct locations to fill the containers almost to the top. Add a few drops of dish washing detergent to separate the components of soil. Label each container based on the location where the soil was excavated.

  ••• **NOTE:** Mineral soil from the ground will separate in the jar. If your garden has potting soil this is not actually real mineral soil. Therefore, it will not separate in the jar as well as true mineral soil. This will serve as an example of the difference between potting soil an true mineral soil.

- Each student should vigorously shake each container.
- Place the containers in a highly visible location where they can be observed for several days.

EXPLANATION:

- **DAY 2-4:** Allow the container to sit for two to three days in order for the contents to settle.
- During days two and three guide students through the completion of the *Shake Up That Soil* student investigation guide. Pose the questions below.
  What do you notice about the samples?
  How are they the same?
  How are they different?
  What is happening in the containers?
- Explain that soil is made up of many living and non-living ingredients including organic matter decaying plants and animals , air, water and mineral matter sand, silt, and clay. In the jar, students will be observing the mineral material in the soil.
- Explain that sand particles are both the largest and the heaviest particles. These are the particles that have settled at the bottom of the sample.
- Explain that silt is the medium size particle and will likely have settled in the middle.
- Explain that clay is the smallest particle.
- Some particles are so small that they may actually float on top of the water.
- Vocabulary: *sand*, *silt*, *clay*, *soil*, *organic material*, *separate*.

ELABORATION:

- Upon completion of the *Shake Up That Soil* student investigation guide, teacher will encourage students to share their observations.
- Teacher will use the questions below to conduct a classroom discussion in order for students to apply their knowledge of the components of soil.
  –Are some soils better for growing plants?
  –Can we add things to our soil to make it better for growing plants?
  –Are there living things that make a habitat in the soil?
  –Can these living things help plants to grow?
  –Can these living things harm plants?

EVALUATION:

- Students will turn in the *Shake Up That Soil* student investigation guide.
- Teacher will observe reactions and interactions with other students and the soil samples.

SHAKE UP THAT SOIL

Name

Draw a picture of the soil in the jar.

Day 1	Day 2
Day 3	Day 4

Circle the parts of soil.

Sand Milk Clay Wood Silt
Wood

KINDERGARTEN LESSON PLAN
Tools Are Not Toys

SUBJECT / GRADE LEVEL: Kindergarten - Inquiry and Exploration - Gardening Safety.

ACADEMIC STANDARDS AND PERFORMANCE INDICATORS:

ELA:

RI.MC.5.1 With guidance and support, ask and answer who, what, when, where, why, and how questions about a text; refer to key details to make inferences and draw conclusions in texts heard or read.

RI.LCS.9.1 With guidance and support, ask and answer questions about known and unknown words.

RI.LCS.8.2 With guidance and support, use front cover, title page, illustrations/ photographs, fonts, glossary, and table of contents to locate and describe key facts or information; describe the relationship between these features and the text.

RL.LCS.10.1 With guidance and support, ask and answer questions about known and unknown words.

RL.RC.13.1 Engage in whole and small group reading with purpose and understanding.

W.MCC.2.1 Use a combination of drawing, dictating, and writing to compose informative/explanatory texts that name and supply information about the topic.

W.MCC.2.2 With guidance and support, plan, revise, and edit building on personal ideas and the ideas of others to strengthen writing.

C.MC.1.4 Participate in conversations with varied partners about focused grade level topics and texts in small and large groups.

C.MC.2.1 With guidance and support, recall information from experiences or gather information from sources to ask and answer questions.

C.MC.2.2 With guidance and support, participate in shared research exploring a variety of texts; express opinions and talk about findings.

C.MC. 3.2 Use appropriate props, images, or illustrations to support verbal communication.

LESSON OBJECTIVES:

TSW practice gardening safety procedures. TSW use tools in an appropriate, safe manner.

Be assisted by the teacher, assistant, shadow, or another student.

MATERIALS:

- Trowel or shovel
- Magnifying glass or lens
- Rake
- Watering can
- Gloves
- *Tools Are Not Toys* student investigation guide (attached)
- *Tools Are Not Toys* teacher guide (attached)

ENGAGEMENT:

- Teacher will hold up a hand trowel and pose the question, "**Is this a toy?**"
- Teacher will follow the engagement question by reading *Jack's Garden* by Henry Cole to the class.

EXPLORATION:

- Teacher will use a series of questions to conduct a classroom discussion in order to allow students to analyze and compare different tools used in the garden.

EXPLANATION:

- Vocabulary: shovel, trowel, rake, glove, apron, watering can.
- Teacher will conduct a charades type game where students pretend to be safely using a garden tool. The students must guess which tool is being used and then explain how to use that tool safely.
- Teacher will explain the vocabulary after completion of the game.

ELABORATION:

- Students will practice their illustration skills while completing the *Tools Are Not Toys* student investigation guide.
- Upon completion, the teacher will encourage students to share their drawings.

EVALUATION:

- Students will turn in the *Tools Are Not Toys* student investigation guide.
- Teacher will observe reactions and interactions with other students and the tools.

RESOURCES:

1. *Jack's Garden* by Henry Cole.
2. *Tools Are Not Toys* handout (attached).

TEACHER GUIDE
TOOLS ARE NOT TOYS

BACKGROUND:

Students should understand that tools make gardening jobs easier but they must be used safely. Students should identify different tools and the procedures for using those tools in an appropriate, safe manner.

ENGAGEMENT:

- Teacher will hold up a hand trowel and pose the question, "**Is this a toy?**"
- Teacher will follow the engagement question by reading *Jack's Garden* by Henry Cole to the class.

EXPLORATION:

- Teacher will use the questions below to conduct a classroom discussion to allow students to analyze and compare different tools used in the garden.
  –Talk about the proper names for tools.
  –How is each tool used?
  –Discuss how each tool looks, and explain how it is used.
  –Discuss how to safely use each tool.
  –Practice how to properly clean and store tools.

EXPLANATION:

- Teacher will conduct a charades type game where students pretend to be safely using a garden tool. The students must guess which tool is being used and then explain how to use that tool safely.
- Vocabulary: *shovel*, *trowel*, *rake*, *glove*, *apron*, *watering can*.

ELABORATION:

- Students will complete the *Tools Are Not Toys* student investigation guide. Upon completion, the teacher will encourage students to share their drawings.
- Assign the position of "tool leader" to a student. He or she will ensure that the class follows proper procedures each time the students work in the garden.

EVALUATION:

- Students will turn in the student *Tools Are Not Toys* investigation guide.
- Teacher will observe reactions and interactions with other students and the tools.

TOOLS ARE NOT TOYS

Name

Draw a picture of yourself using a garden tool. Write a sentence about the tool.

KINDERGARTEN LESSON PLAN
Measuring Temperature

SUBJECT / GRADE LEVEL: Kindergarten - Earth Science.

ACADEMIC STANDARDS AND PERFORMANCE INDICATORS:

Science:

K.E.3A.1 Analyze and interpret local weather condition data including precipitation, wind, temperature, and cloud cover to describe weather patterns that occur from day to day, using simple graphs and pictorial weather symbols.

LESSON OBJECTIVES:

TSW will practice reading thermometers and discover how temperature affects the garden.

MATERIALS:

- Multiple thermometers amount varies depending on the size of the class
- *Measuring Temperature* student investigation guide (attached)
- *Measuring Temperature* teacher guide (attached)

ENGAGEMENT:

- Teacher will pose the question, "**Describe the temperature outside today?**"
- Ask students to share words that describe temperature, such as hot, warm, cold, chilly, etc...

EXPLORATION:

- Teacher will introduce focus questions and guide students through methods to observe and record data.

EXPLANATION:

- Teacher will explain the vocabulary while students investigate weather patterns through inquiry based questioning.

ELABORATION:

- Teacher will gather students together as a group once and allow them to share what they noticed and learned.
- Students will chart their data on the *Measuring Temperature* student investigation guide.

EVALUATION:

- Students will turn in the *Measuring Temperature* student investigation guide.
- Teacher will observe reactions and interactions with other students and the thermometers.
- Students will be assessed on their ability to read thermometers through one-on-one conferencing with the teacher.

RESOURCES:

1. *Measuring Temperature* handout (attached).

TEACHER GUIDE

MEASURING TEMPERATURE

BACKGROUND:

Students should understand that temperature can positively and negatively affect plant growth.

ENGAGEMENT:

- Teacher will pose the question, "**Describe the temperature outside today?**"
- Ask students to share words that describe temperature, such as hot, warm, cold, chilly, etc...

EXPLORATION:

- Introduce the focus questions:
  –**How can we measure temperature in the garden?**
  –**Why do gardeners need to keep track of the temperature?**
- Show the children the thermometer.
- Ask them to describe it. **Have they ever used a thermometer before?**
- Review ways in which the class can observe and record data.
- Place a few thermometers in a sunny location and place a few in the shade. Let the children discuss what they think will happen to the thermometers in the different temperatures.
- Divide into small groups to observe the different thermometers. Demonstrate how to read the thermometers and record the findings.

EXPLANATION:

- Gather students together as a group and allow them to share what they noticed and learned. **What did you notice about the temperature in the shade compared to the temperature in the sun?**
- Next, ask students why they think gardeners need to know the temperature. **How does temperature affect the garden?**
- Vocabulary: *temperature*, *measure*, *thermometer*, *tool*, *shade*, *mercury*.

ELABORATION:

- Over the course of a week, guide students through taking a recording of the temperature in the garden. Choose several times per day such as: Start of school / Before lunch / After recess.
- Students will chart their data on the *Measuring Temperature* student investigation guide.

- Finally, students will analyze the temperature patterns for the week looking at temperatures and how they relate to sunny days, cloudy days and rainy days.

EVALUATION:

- Students will turn in the *Measuring Temperature* student investigation guide.
- Teacher will observe reactions and interactions with other students and the thermometers.
- Students will be assessed on their ability to read thermometers through one-on-one conferencing with the teacher.

MEASURING TEMPERATURE

Name

Write one word to describe each weather picture.

Write the temperature for each day and draw one of the above symbols to describe the weather.

Monday	Tuesday	Wednesday	Thursday	Friday

KINDERGARTEN LESSON PLAN
Liking Leaves

SUBJECT / GRADE LEVEL: Kindergarten - Exploring the Observable Properties of Matter.

ACADEMIC STANDARDS AND PERFORMANCE INDICATORS:

ELA:

RI.MC.5.1 With guidance and support, ask and answer who, what, when, where, why, and how questions about a text; refer to key details to make inferences and draw conclusions in texts heard or read.

RI.LCS.9.1 With guidance and support, ask and answer questions about known and unknown words.

RI.LCS.8.2 With guidance and support, use front cover, title page, illustrations/ photographs, fonts, glossary, and table of contents to locate and describe key facts or information; describe the relationship between these features and the text.

RL.LCS.10.1 With guidance and support, ask and answer questions about known and unknown words.

RL.RC.13.1 Engage in whole and small group reading with purpose and understanding.

W.MCC.2.1 Use a combination of drawing, dictating, and writing to compose informative/explanatory texts that name and supply information about the topic.

W.MCC.2.2 With guidance and support, plan, revise, and edit building on personal ideas and the ideas of others to strengthen writing.

C.MC.1.4 Participate in conversations with varied partners about focused grade level topics and texts in small and large groups.

C.MC.2.1 With guidance and support, recall information from experiences or gather information from sources to ask and answer questions.

C.MC.2.2 With guidance and support, participate in shared research exploring a variety of texts; express opinions and talk about findings.

C.MC. 3.2 Use appropriate props, images, or illustrations to support verbal communication.

Science:

K.P.4A.1 Analyze and interpret data to compare the qualitative properties of objects such as size, shape, color, texture, weight, flexibility, attraction to magnets, or ability to sink or float, and classify objects based on similar properties.

K.P.4A.2 Develop and use models to describe and compare the properties of different materials including wood, plastic, metal, cloth, and paper and classify materials by their observable properties, by their uses, and whether they are natural or human-made.

Math:

K.MDA.3 Sort and classify data into 2 or 3 categories with data not to exceed 20 items in each category.

K.G.1 Describe positions of objects by appropriately using terms, including below, above, beside, between, inside, outside, in front of, or behind.

K.G.2 Identify and describe a given shape and shapes of objects in everyday situations to include two-dimensional shapes i.e., triangle, square, rectangle, hexagon, and circle and three-dimensional shapes i.e., cone, cube, cylinder, and sphere.

K.G.4 Analyze and compare two- and three-dimensional shapes of different sizes and orientations using informal language.

LESSON OBJECTIVES:

TSW examine leaves to observe different characteristics. TSW sort leaves based on physical characteristics.

MATERIALS:

- Leaves
- Paper
- Pencils/marker
- Paper plates
- *Leaf Rides the Wind* by Celeste Mannis
- *Leaf Man* by Lois Ehlert
- *Liking Leaves* student illustration exercise (attached)
- *Liking Leaves* teacher guide (attached)

ENGAGEMENT:

- Teacher will pose the question, "**Do all leaves look the same?**"
- Students will describe the shapes of different leaves that they have seen.
- Teacher will follow the engagement question by reading *Leaf Rides the Wind* by Celeste Mannis or *Leaf Man* by Lois Ehlert to the class.

EXPLORATION:

- Teacher will guide students through an exercise in which they will collect, examine and sort leaves.

EXPLANATION:

- Teacher will explain the importance of leaves while students investigate the characteristics of leaves through inquiry based questioning.

ELABORATION:

- Students will practice their observation and illustration skills while completing their *Liking Leaves* student illustration exercise.
- Teacher will conduct a classroom discussion in order for students to apply their leaf knowledge.

EVALUATION:

- Students will turn in their *Liking Leaves* student illustration exercise.
- Teacher will observe reactions and interactions with other students and the leaves.

RESOURCES:

1. *Leaf Rides the Wind* by Celeste Mannis.
2. *Leaf Man* by Lois Ehlert.
3. *Liking Leaves* handout (attached).

TEACHER GUIDE
LIKING LEAVES

BACKGROUND:

Students should understand that leaves come in many shapes and sizes which allow them to conduct their very important job of collecting sunlight for photosynthesis.

ENGAGEMENT:

- Teacher will pose the question, "**Do all leaves look the same?**"
- Students will describe the shapes of different leaves that they have seen.
- Teacher will follow the engagement question by reading *Leaf Rides the Wind* by Celeste Mannis or *Leaf Man* by Lois Ehlert to the class.

EXPLORATION:

- Students will go out in the garden or other outdoor space and collect three different leaves.
- Students should examine the texture, color, size, and shape of each leaf.
- Gather students around a large table and have them pile the leaves on top of the table.
- Invite the students to sort the leaves onto different paper plates. Each plate is one category.
- **Why did you sort the leaves the way that you did?**

EXPLANATION:

- Leaves are important because they collect sunlight which the plant converts to energy. That energy allows the plant to be healthy and grow strong.
- **Why are leaves shaped differently?**
- **Are some leaves dangerous?**
- **What might happen to a plant if it did not have any leaves?**
- **Why do leaves change color?**
- **Why do some trees keep their leaves and others lose their leaves?**

ELABORATION:

- Students will complete the *Liking Leaves* student illustration.
- Teacher will encourage students to share their illustrations. Students will use a leaf to create a picture. For example, one illustration may use the leaf as the body of a ladybug.
- Other extension exercises include:

–Have students sort and classify other objects in the classroom.

–Create leaf rubbings.

–Create a leaf man based on reading.

–Have students describe the leaves using terms for their qualities and properties, including color, size, texture, and smell.

–Have students discuss and evaluate the criteria other students used to sort their leaves.

EVALUATION:

- Students will turn in their *Liking Leaves* student illustration.
- Teacher will observe reactions and interactions with other students and the leaves.

LIKING LEAVES

Name

____

Glue a leaf in the box. Use your leaf to create a picture.

My leaf has become a

____.

KINDERGARTEN LESSON PLAN
Look Up, Look Down, Look Around

SUBJECT / GRADE LEVEL: Kindergarten - Exploring Organisms and the Environment.

ACADEMIC STANDARDS AND PERFORMANCE INDICATORS:

ELA:

RL.MC.5.2 With guidance and support, ask and answer questions to make predictions using prior knowledge, pictures, illustrations, title, and information about author and illustrator.

C.MC.1.4 Participate in conversations with varied partners about focused grade level topics and texts in small and large groups.

Science:

K.S1.A.3 With teacher guidance, conduct structured investigations to answer scientific questions, test predictions and develop explanations: 1.) predict possible outcomes, 2.) identify materials and follow procedures, 3.) use appropriate tools or instruments to make qualitative observations and take nonstandard measurements, and 4.) record and represent data in an appropriate form. Use appropriate safety procedures.

LESSON OBJECTIVES:

TSW will locate, identify and observe animals in their environment.

MATERIALS:

- Magnifying glasses
- Soil with a worm
- Flowers with a small bug
- *Streganona's Harvest* by Tomie dePaola
- *Why Animals Live in Burrows* by Valerie Weber
- *Animals Under Our Feet* by Sindy McKay
- *Look Up, Look Down, Look Around* student investigation guide (attached)
- *Look Up, Look Down, Look Around* teacher guide (attached)

ENGAGEMENT:

- Teacher will pose the question, "**Are there animals under the ground?**"

- Teacher will follow the engagement question by reading *Why Animals Live in Burrows* by Valerie Weber or *Animals Under Our Feet* by Sindy McKay.

EXPLORATION:

- Teacher will guide students through the garden to look for signs of wildlife.

EXPLANATION:

- Teacher will explain the vocabulary while students investigate animal habitats and answer inquiry based questions.

ELABORATION:

- Students will practice their observation skills while completing the *Look Up, Look Down, Look Around* investigation guide.

EVALUATION:

- Students will turn in the *Look Up, Look Down, Look Around* student investigation guide.
- Teacher will observe reactions and interactions with other students and the animals.

RESOURCES:

1. *Streganona's Harvest* by Tomie dePaola.
2. *Why Animals Live in Burrows?* by Valerie Weber.
3. *Animals Under Our Feet* by Sindy McKay.
4. *Look Up, Look Down, Look Around* handout (attached).

TEACHER GUIDE

LOOK UP, LOOK DOWN, LOOK AROUND

BACKGROUND:

Students should understand that the garden provides a habitat for many animals both above and below the ground.

ENGAGEMENT:

- Teacher will pose the question, "**Are there animals under the ground?**"
- Teacher will follow the engagement question by reading *Why Animals Live in Burrows* by Valerie Weber or *Animals Under Our Feet* by Sindy McKay.

EXPLORATION:

- Students will go out into the garden and look for signs of wildlife. Ask students to describe the signs of wildlife. Pose the questions below.
  –Do you think there are any animals living underground?
  –Where do these animals go during the day?
  –Are there plants in the garden that serve as food or habitat for wildlife?
  –What do you think would be under a rock?
  –What evidence can you find that animals, tiny or large, coexist with you?

EXPLANATION:

- Explain to students that flowers need pollinators such as bees, butterflies, and bats. Lots of animals coexist with us, even if we don't always see them. Some are very tiny. Some live beneath the ground in burrows.
- Gently look under a rock or a fallen tree to discover what is living beneath it. Even though we might not see an animal, if we find evidence, the animal is most likely still nearby.
  A burrow's entrance might house a mouse that sleeps during the day.
  A cracked nutshell may have been dropped by a bird who was having lunch and then flew off in search of more food.
- Vocabulary: *coexist*, *exist*, *insect*, *animal*, *burrow*, *nibbled leaf.*

ELABORATION:

- Possible extension activities include:
  –Construct a toad house and place it in the garden.
- Read *Streganona's Harvest* by Tomie dePaola.
- Students will illustrate their findings in the *Look Up, Look Down, Look Around* student investigation guide.

EVALUATION:

- Students will turn in the *Look Up, Look Down, Look Around* student investigation guide.
- Teacher will observe reactions and interactions with other students and the animals.

LOOK UP, LOOK DOWN, LOOK AROUND

Name

Draw a picture of four animals that you would find in the garden.

On a plant	**In the air**
Underground	**Above ground**

Circle the animals that you would find in a garden.

Grasshopper   Crab   Rabbit   Ladybug   Fish   Worm

KINDERGARTEN LESSON PLAN

Ready, Set, Roots

SUBJECT / GRADE LEVEL: Kindergarten - Exploring Organisms and the Environment.

ACADEMIC STANDARDS AND PERFORMANCE INDICATORS:

ELA:

RI.MC.5.1 With guidance and support, ask and answer who, what, when, where, why, and how questions about a text; refer to key details to make inferences and draw conclusions in texts heard or read.

RI.LCS.9.1 With guidance and support, ask and answer questions about known and unknown words.

RI.LCS.8.2 With guidance and support, use front cover, title page, illustrations/ photographs, fonts, glossary, and table of contents to locate and describe key facts or information; describe the relationship between these features and the text.

RL.LCS.10.1 With guidance and support, ask and answer questions about known and unknown words.

RL.RC.13.1 Engage in whole and small group reading with purpose and understanding.

W.MCC.2.1 Use a combination of drawing, dictating, and writing to compose informative/explanatory texts that name and supply information about the topic.

W.MCC.2.2 With guidance and support, plan, revise, and edit building on personal ideas and the ideas of others to strengthen writing.

C.MC.1.4 Participate in conversations with varied partners about focused grade level topics and texts in small and large groups.

C.MC.2.1 With guidance and support, recall information from experiences or gather information from sources to ask and answer questions.

C.MC.2.2 With guidance and support, participate in shared research exploring a variety of texts; express opinions and talk about findings.

C.MC. 3.2 Use appropriate props, images, or illustrations to support verbal communication.

Science:

K.L.2A.1 Obtain information to answer questions about different organisms found in the environment such as plants, animals, and fungi.

K.L.2A.4 Analyze and interpret data to describe how humans use their senses to learn about the world around them.

Math:

K.MDA.1 Identify measurable attributes length, weight of an object.

LESSON OBJECTIVES:

TSW identify what plants need to survive.

MATERIALS:

- Lima bean seeds
- Cups
- Soil
- Window or an area with sunlight
- Water
- *Grow Flower Grow* by Lisa Bruce
- *Ready, Set, Roots* student investigation guide (attached)
- *Ready, Set, Roots* teacher guide (attached)

ENGAGEMENT:

- Teacher will pose the question, "**What do plants need to survive?**"
- Teacher will follow the engagement question by reading *Grow Flower Grow* by Lisa Bruce to the class.

EXPLORATION:

- Teacher will guide students through an exercise in which they plant and observe two lima bean seeds.

EXPLANATION:

- Teacher will explain the vocabulary while students investigate needs of plants with inquiry based questioning.

ELABORATION:

- Students will practice their observation skills while completing the R*eady, Set, Roots* student investigation guide.
- Teacher will conduct a classroom discussion in order for students to apply their knowledge of plant growth.

EVALUATION:

- Students will turn in the *Ready, Set, Roots* student investigation guide.
- Teacher will observe reactions and interactions with other students and the lima bean plants.

RESOURCES:

1. *Grow Flower Grow* by Lisa Bruce.
2. *Ready, Set, Roots* handout (attached).

TEACHER GUIDE
READY, SET, ROOTS

BACKGROUND:

Students should understand that plants need light and water to survive and grow. Students should identify the roots of a plant and understand the important role that they play.

ENGAGEMENT:

- Teacher will pose the question, "**What do plants need to survive?**"
- Teacher will follow the engagement question by reading *Grow Flower Grow* by Lisa Bruce to the class.

EXPLORATION:

- Give each student two lima bean seeds.
- Students should plant each lima bean seed in a cup and place both cups in a well-lit area, such as a window.
- Over the course of ten days, students will water plant A, but not plant B.

EXPLANATION:

- Explain to the students that just like people, plants need water to be healthy and grow. The roots suck up water from the soil like a straw. Since plant B did not have much water, it did not grow as well as plant A. Pose the following questions.
  –What would happen if we watered both plants, but only put one in the sunlight?
  –What is happening with each plant?
  –Why do you think plant A is growing faster than plant B?
  –Is there anything we could have done differently to help plant B grow?
- Vocabulary: *roots*, *stem*, *sunlight*, *water*, *survive*.

ELABORATION:

- Students should measure the height of plant A and plant B each day and record the results on their *Ready, Set, Roots* student investigation guide. Students will complete the Venn Diagram on the *Ready, Set, Roots* student investigation guide.

- Upon completion of the *Ready, Set, Roots* student investigation guide, teacher will encourage students to share their observations.
- Teacher will use the questions below to conduct a classroom discussion in order for students to apply their knowledge of plant growth.
  –**What do you think plants need to survive?**
  –**What do we need to survive?**
  –**Do you think we need the same things as plants?**

EVALUATION:

- Students will turn in the *Ready, Set, Roots* student investigation guide.
- Teacher will observe reactions and interactions with other students and the lima bean plants.

READY, SET, ROOTS

Name ____

Record the plant height for each day.

Day 1	Day 2	Day 3	Day 4	Day 5	Day 6	Day 7	Day 8	Day 9	Day 10

List three needs of both plants and people. List one need of both.

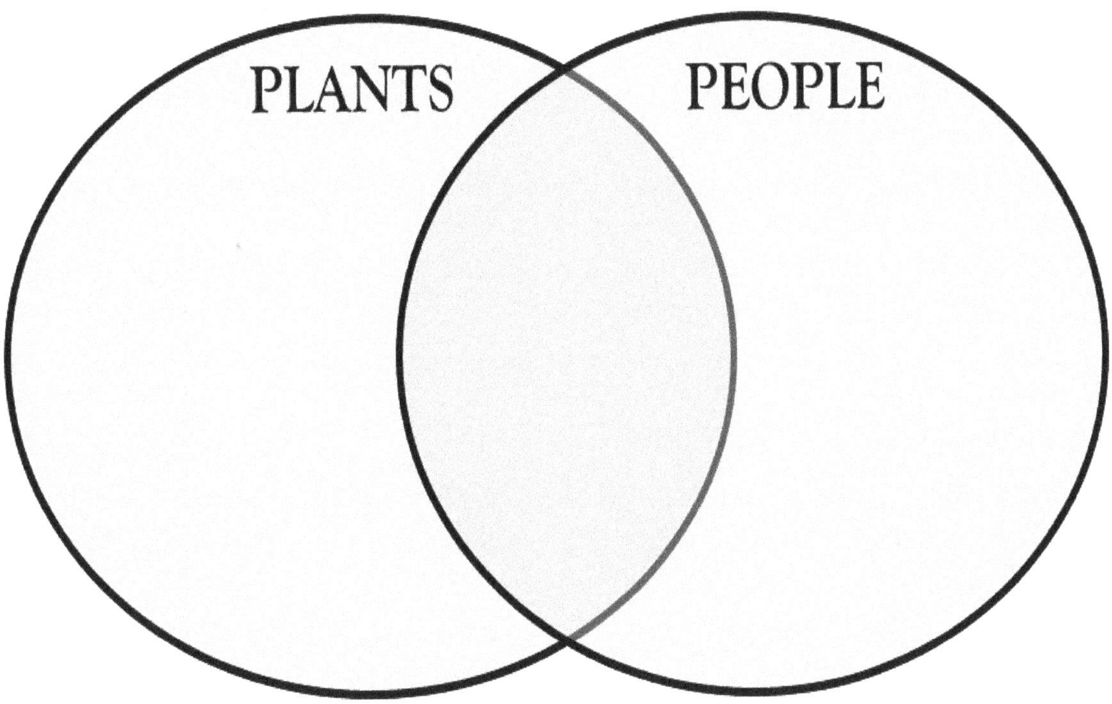

KINDERGARTEN LESSON PLAN 10

Roving Rain Drops

SUBJECT / GRADE LEVEL: Kindergarten - Exploring Weather Patterns.

ACADEMIC STANDARDS AND PERFORMANCE INDICATORS:

ELA:

RI.MC.5.1 With guidance and support, ask and answer who, what, when, where, why, and how questions about a text; refer to key details to make inferences and draw conclusions in texts heard or read.

RI.LCS.9.1 With guidance and support, ask and answer questions about known and unknown words.

RI.LCS.8.2 With guidance and support, use front cover, title page, illustrations/ photographs, fonts, glossary, and table of contents to locate and describe key facts or information; describe the relationship between these features and the text.

RL.LCS.10.1 With guidance and support, ask and answer questions about known and unknown words.

RL.RC.13.1 Engage in whole and small group reading with purpose and understanding.

W.MCC.2.1 Use a combination of drawing, dictating, and writing to compose informative/explanatory texts that name and supply information about the topic.

W.MCC.2.2 With guidance and support, plan, revise, and edit building on personal ideas and the ideas of others to strengthen writing.

C.MC.1.4 Participate in conversations with varied partners about focused grade level topics and texts in small and large groups.

C.MC.2.1 With guidance and support, recall information from experiences or gather information from sources to ask and answer questions.

C.MC.2.2 With guidance and support, participate in shared research exploring a variety of texts; express opinions and talk about findings.

C.MC. 3.2 Use appropriate props, images, or illustrations to support verbal communication.

Science:

K.E.3A.1 Analyze and interpret local weather condition data including precipitation, wind, temperature, and cloud cover to describe weather patterns that occur from day to day, using simple graphs and pictorial weather symbols.

K.E.3A.2 Develop and use models to predict seasonal weather patterns and changes.

K.E.3A.4 Define problems caused by the effects of weather on human activities and design solutions or devices to solve the problem.

Math:

K.NS.1 Count forward by ones and tens to 100.

K.MDA.2 Compare objects using words such as shorter/longer, shorter/taller, and lighter/heavier.

K.MDA.3 Sort and classify data into 2 or 3 categories with data not to exceed 20 items in each category.

LESSON OBJECTIVES:

TSW will make a rain gauge to measure how much rain falls in the garden, graph results, and describe weather patterns.

MATERIALS:

- Ruler
- Clear, wide-mouthed jar
- Paper
- Pencil
- Tape
- *The Water Journey* by Eleanor Schmid
- *Water* by Frank Ash
- *Roving Raindrops* student investigation guide (attached)
- *Roving Raindrops* teacher guide (attached)

ENGAGEMENT:

- Teacher will pose the question, "**Where does rain go after it falls from the sky?**"
- Teacher will follow the engagement question by reading *The Water Journey* by Eleanor Schmid or *Water* by Frank Ash to the class.

EXPLORATION:

- Teacher will guide students through an exercise in which they will construct a rain gauge.

EXPLANATION:

- The class will record and discuss how much rain has fallen in the jar in a specific period of time such as daily or weekly.

ELABORATION:

- Students will complete the *Roving Raindrops* student investigation guide. Teacher will encourage students to share their observations.

EVALUATION:

- The class will graph the daily recordings of rainfall from the rain gauge.
- Students will turn in the *Roving Raindrops* student investigation guide.
- Teacher will observe reactions and interactions with other students and the rain gauge.

RESOURCES:

1. *The Water Journey* by Eleanor Schmid.
2. *Water* by Frank Ash.
3. *Roving Raindrops* handout (attached).

TEACHER GUIDE
ROVING RAINDROPS

BACKGROUND:

Students should understand that raindrops, during their journey, provide water to plants as well as humans.

ENGAGEMENT:

- Teacher will pose the question, "**Where does rain go after it falls from the sky?**"
- Teacher will follow the engagement question by reading *The Water Journey* by Eleanor Schmid or *Water* by Frank Ash to the class.

EXPLORATION:

Teacher will lead the class in the construction of a rain gauge using the following steps:

- Place a ruler in the jar of the rain gauge. Tape the ruler to the edge of the jar with the numbers facing outward and the smallest numbers toward the bottom.
- Take your jar outside and place it in an open area. Make sure it is not under any trees, but also try to put it somewhere where it will not get turned over.
- After the first rain, go out to your rain gauge and read the ruler to measure how high the water rose.
- Record data on class chart.

EXPLANATION:

- The class will record and discuss how much rain has fallen in the jar in a specific period of time such as daily or weekly.
- Teacher will use the questions below to conduct a classroom discussion in order for students to investigate rain.

  –How do we measure rainfall?
  –Where does the rain come from?
  –Why do we need rain? What would happen if we didn't have rain?
  –Can you live without water? What about plants?

ELABORATION:

- Students will complete the *Roving Raindrops* student investigation guide, teacher will encourage students to share their observations.
- Teacher can implement these additional extension activities if time allows:
  –Students can check the local weather forecast and see how the rain level in the garden compares to the amount of rain that fell in other areas near the school.
  –Students use the water in the jar to water some plants.
  –Set up a water table to explore the properties of water using the senses.

EVALUATION:

- The class will graph the daily recordings of rainfall from the rain gauge.
- Students will turn in the *Roving Raindrops* student investigation guide.
- Teacher will observe reactions and interactions with other students and the rain gauge.

ROVING RAINDROPS

Name

Draw a picture of a raindrop and the journey it travels. Write a sentence below to describe your picture.

KINDERGARTEN LESSON PLAN

Weather or Not Weather Changes Through the Seasons

SUBJECT / GRADE LEVEL: Kindergarten - Exploring Weather Patterns.

ACADEMIC STANDARDS AND PERFORMANCE INDICATORS:

ELA:

RI.MC.5.1 With guidance and support, ask and answer who, what, when, where, why, and how questions about a text; refer to key details to make inferences and draw conclusions in texts heard or read.

RI.LCS.9.1 With guidance and support, ask and answer questions about known and unknown words.

RI.LCS.8.2 With guidance and support, use front cover, title page, illustrations/ photographs, fonts, glossary, and table of contents to locate and describe key facts or information; describe the relationship between these features and the text.

RL.LCS.10.1 With guidance and support, ask and answer questions about known and unknown words.

RL.RC.13.1 Engage in whole and small group reading with purpose and understanding.

W.MCC.2.1 Use a combination of drawing, dictating, and writing to compose informative/explanatory texts that name and supply information about the topic.

W.MCC.2.2 With guidance and support, plan, revise, and edit building on personal ideas and the ideas of others to strengthen writing.

C.MC.1.4 Participate in conversations with varied partners about focused grade level topics and texts in small and large groups.

C.MC.2.1 With guidance and support, recall information from experiences or gather information from sources to ask and answer questions.

C.MC.2.2 With guidance and support, participate in shared research exploring a variety of texts; express opinions and talk about findings.

C.MC.3.2 Use appropriate props, images, or illustrations to support verbal communication.

Science:

K.E.3A.3 Obtain and communicate information to support claims about how changes in seasons affect plant and animals.

K.E.3A.4 Define problems caused by the effects of weather on human activities and design solutions or devices to solve the problem.

Math:

K.MDA.3 Sort and classify data into 2 or 3 categories with data not to exceed 20 items in each category.

LESSON OBJECTIVES:

TSW collect information from a daily weather tracking report. TSW explain how the change of weather through the season affects the plants and animals in the garden.

MATERIALS:

- Blank calendar for each month
- Large class calendar
- Camera or iPad with camera function
- *Four Stories for Four Seasons* by Tomie De Paola
- *The Reason for the Seasons* by Gail Gibbons
- *What Makes the Seasons?* by Megan Cash
- *Shake Up That Soil* student investigation guide (attached)
- *Shake Up That Soil* teacher guide (attached)

ENGAGEMENT:

- Teacher will pose the question, "**What do you know about the four seasons?**"
- Teacher will follow the engagement question by reading *Four Stories for Four Seasons* by Tomie De Paola, *The Reason for the Seasons* by Gail Gibbons or *What Makes the Seasons?* by Megan Cash to the class.

EXPLORATION:

- Teacher will guide students through the selection of a person that will report the weather to the class.
- Teacher will discuss the weather documentation using the monthly calendar.

EXPLANATION:

- Teacher will take the class to the garden.
- Teacher will use the questions below to conduct a class discussion in order for students to investigate the relationship between plants and the weather.

ELABORATION:

- Students will practice their illustration skills while completing the *Weather or Not* student investigation guide.
- Teacher will conduct a classroom discussion in order for students to apply their knowledge of the weather.

EVALUATION:

- Students will turn in the *Weather or Not* student investigation guide.

RESOURCES:

1. *Four Stories for Four Seasons* by Tomie De Paola.
2. *The Reason for the Seasons* by Gail Gibbons.
3. *What Makes the Seasons?* by Megan Cash.
4. *Weather or Not* handout (attached).

TEACHER GUIDE

WEATHER OR NOT

BACKGROUND:

Students should understand that plant growth can be positively and negatively impacted by the weather.

ENGAGEMENT:

- Teacher will pose the question, "**What do you know about the four seasons?**"
- Teacher will follow the engagement question by reading *Four Stories for Four Seasons* by Tomie De Paola, *The Reason for the Seasons* by Gail Gibbons or *What Makes the Seasons?* by Megan Cash to the class.

EXPLORATION:

- The class or teacher chooses a weather man or woman for the day or week.
- With teacher assistance, the weather person can take a picture of the garden at the beginning of each day.
- The weather person can report the weather to the class, and put a picture of the weather on the class calendar.
- Each student can record the weather in their science journal by illustrating what they see.
- The teacher and the students will discuss how weather changes affect plants, animals, and people.
- What do you notice about the weather each day?

EXPLANATION:

- Teacher will take the class to the garden. Teacher will use the questions below to conduct a class discussion in order for students to investigate the relationship between plants and the weather.
  –How do some plants or trees change during the seasons?
  –How do some animals behave differently during the weather changes?
  –How are people affected by seasonal weather changes?
  –How do the weather changes affect our garden?
  –If it rains outside, do we still need to water our plants?
- Vocabulary: *windy, cloudy, rainy, sunny, partly cloudy, snowy, foggy.*

ELABORATION:

- Students can compile the daily reports and make a book or movie about the weather through the course of the year.
- Students can research weather and its effects on a specific plant, animal, or place.
- Throughout the school year, teacher should revisit the weather and seasons as they change.

EVALUATION:

- Teacher can examine students weather observations in their science journals and have discussions about the weather.
- Teacher should observe students interacting during classroom activities and discussions about weather.
- Students will turn in the *Weather or Not* student investigation guide.

WEATHER OR NOT

Name

Draw a picture of the weather that goes with each situation below.

Plants grow well.	Plants need water.
Plants grow slowly.	Plants don't need water.

What is your favorite season?

KINDERGARTEN LESSON PLAN
What is a Weed?

SUBJECT / GRADE LEVEL: Kindergarten - Exploring Organisms in the Environment.

ACADEMIC STANDARDS AND PERFORMANCE INDICATORS:

ELA:

RI.MC.5.1 With guidance and support, ask and answer who, what, when, where, why, and how questions about a text; refer to key details to make inferences and draw conclusions in texts heard or read.

RI.LCS.9.1 With guidance and support, ask and answer questions about known and unknown words.

RI.LCS.8.2 With guidance and support, use front cover, title page, illustrations/ photographs, fonts, glossary, and table of contents to locate and describe key facts or information; describe the relationship between these features and the text.

RL.LCS.10.1 With guidance and support, ask and answer questions about known and unknown words.

RL.RC.13.1 Engage in whole and small group reading with purpose and understanding.

W.MCC.2.1 Use a combination of drawing, dictating, and writing to compose informative/explanatory texts that name and supply information about the topic.

W.MCC.2.2 With guidance and support, plan, revise, and edit building on personal ideas and the ideas of others to strengthen writing.

C.MC.1.4 Participate in conversations with varied partners about focused grade level topics and texts in small and large groups.

C.MC.2.1 With guidance and support, recall information from experiences or gather information from sources to ask and answer questions.

C.MC.2.2 With guidance and support, participate in shared research exploring a variety of texts; express opinions and talk about findings.

C.MC. 3.2 Use appropriate props, images, or illustrations to support verbal communication.

Science:

K.L.2A.1 Obtain information to answer questions about different organisms found in the environment such as plants, animals, and fungi.

LESSON OBJECTIVES:

TSW determine the difference between a weed and a plant or flower in the garden.

MATERIALS:

- *Flower Garden* by Henry Cole
- Glue
- *What is a Weed?* student illustration exercise
- *What is a Weed?* teacher guide (attached)

ENGAGEMENT:

- Teacher will pose the question, "**What is a weed?**"
- Teacher will follow the engagement question by reading *Flower Garden* by Henry Cole to the class.

EXPLORATION:

- Teacher will take students out in the garden.
- Teacher will use inquiry based questions to conduct a class discussion in order for students to investigate the properties of weeds.

EXPLANATION:

- Teacher will explain the vocabulary while students discuss the importance of removing weeds from the garden.

ELABORATION:

- Teacher will conduct a game in which students will locate and dispose of weeds in the garden.
- Students will practice their observation and illustration skills while completing their *What is a Weed?* student illustration exercise.

EVALUATION:

- Students will turn in the *What is a Weed?* student illustration exercise.
- Teacher will observe reactions and interactions with other students and the plants.

RESOURCES:

1. *Flower Garden* by Henry Cole.
2. *What is a Weed* handout (attached).

TEACHER GUIDE
WHAT IS A WEED

BACKGROUND:

Students should understand that weeds compete with other plants for sunlight, water and nutrients.

ENGAGEMENT:

- Teacher will pose the question, "**What is a weed?**"
- Teacher will follow the engagement question by reading *Flower Garden* by Henry Cole to the class.

EXPLORATION:

- Teacher will take students out in the garden. Teacher will use the questions below to conduct a class discussion in order for students to investigate the properties of weeds.
  What is a weed?
  What does a weed look like?
  What do weeds take from other plants?
  Why do we need to remove weeds from the garden?
  How do the weeds look alike?
  How do the weeds look different?

EXPLANATION:

- Teacher will ask students, "**What do weeds take from other plants in the garden?**" Sunlight / Water / Nutrients.
- Students will identify which plants in the garden are weeds.
- Vocabulary: *weed*, *invasive*, *nutrients*.

ELABORATION:

- Play a weed "I Spy" game by having children silently walk through the garden to look for weeds. When a student believes he or she had found a weed, he or she will stop and point until the teacher evaluates if the student is correct. If the student is correct, he or she may pull the weed and remove it from the garden.
- Upon completion of the *What is a Weed?* student illustration, teacher will encourage students to share their illustrations.

EVALUATION:

- Through the weed "I Spy" game, teachers can tell if students are learning to recognize weeds.
- Students will turn in the *What is a Weed?* student investigation guide.
- Teacher will observe reactions and interactions with other students and the plants.

WHAT IS A WEED?

Name

____

Glue a weed in the box. Use your weed to create a picture.

My weed has become a

____.

LESSONS FOR FIRST GRADE

FIRST GRADE LESSON PLAN
Creating Compost

SUBJECT / GRADE LEVEL: First Grade - Earth's Natural Resources.

ACADEMIC STANDARDS AND PERFORMANCE INDICATORS:

ELA :

RI.MC. 5.1 Ask and answer who, what, when, where, why, and how questions about a text; use key details to make inferences and draw conclusions in texts heard or read.

RI.MC.6.1 Retell the central idea and identify key details to summarize a text heard, read, or viewed.

RI.MC.9.1 Ask and answer questions about known and unknown words in a text.

RI.LCS. 8.1 Identify words, phrases, illustrations, and photographs used to provide information.

RI.LCS. 8.2 Use front cover, title page, illustrations/ photographs, fonts, glossary, and table of contents to locate and describe key facts or information; describe the relationship between these features and the text.

W.MCC.1.1 Explore print and multimedia sources to write opinion pieces that introduce the topic, state an opinion, give a reason for the opinion, and provide a sense of closure.

W.MCC.1.2 Plan, revise, and edit building on personal ideas and the ideas of others to strengthen writing.

W.MCC.2.1 Explore print and multimedia sources to write informative/explanatory texts that name a topic, supply facts about the topic, and provide a sense of closure.

Science:

1.E.4B.1 Obtain and communicate information to summarize how natural resources are used in different ways, such as soil and water to grow plants, rocks to make roads, walks, or buildings, or sand to make glass.

1.E.4B.2 Obtain and communicate information to explain ways natural resources can be conserved, such as reducing trash through reuse, recycling, or replanting trees.

LESSON OBJECTIVES:

TSW will investigate the benefits of compost. TSW identify the best materials for creating compost.

MATERIALS:

- Scraps of food *coffee grounds, banana peels, carrot shreds, apple peels. NO citrus, meat, or dairy*
- Compost bin *purchased or built*
- Soil
- 1 yard wide space for the compost bin
- *It's Compost: Growing Gardens From Your Garbage* by Linda Glaser
- *Compost Stew* by Mary McKenna Siddals
- *Creating Compost* student investigation guide (attached)
- *Creating Compost* teacher guide (attached)

ENGAGEMENT:

- Teacher will pose the question, "**What is compost?**"
- Teacher will follow the engagement question by reading *It's Compost: Growing Gardens From Your Garbage* by Linda Glaser or *Compost Stew* by Mary McKenna Siddals to the class.
- Teacher will show the following video to the class: *Riverbanks ComPOOst: It's Heavy Doodie* By Riverbanks Zoo" https://www.youtube.com/watch?v=YnDZ9yAcpnI.

EXPLORATION:

- Teacher will facilitate the setup of the compost bin.
- Teacher will guide students in order to begin filling the compost bin with appropriate materials.

EXPLANATION:

- Teacher will explain the vocabulary while students discuss the benefits of using compost in the garden.

ELABORATION:

- Teacher will take students out to the compost bin. Teacher will use inquiry based questions to conduct a class discussion in order for students to investigate the properties of compost.
- Students will practice their observation and illustration skills while completing their *Creating Compost* student investigation guide.

EVALUATION:

- Students will turn in the *Creating Compost* student investigation guide.
- Teacher will observe reactions and interactions with other students and the compost.

RESOURCES:

1. *It's Compost: Growing Gardens from your Garbage* by Linda Glaser.
2. *Compost Stew* by Mary McKenna Siddals.
3. *Riverbanks ComPOOst: It's Heavy Doodie* By Riverbanks Zoo: https://www.youtube.com/watch?v=YnDZ9yAcpnI.
4. *Creating Compost* handout (attached).

TEACHER GUIDE
CREATING COMPOST

BACKGROUND:

Students should understand that they can use waste from the garden to create nutrient rich compost which can be used to enrich the garden soil and feed the plants.

ENGAGEMENT:

- Teacher will pose the question, "**What is compost?**"
- Teacher will follow the engagement question by reading *It's Compost: Growing Gardens From Your Garbage* by Linda Glaser or *Compost Stew* by Mary McKenna Siddals to the class.
- Teacher will show the following video to the class: *Riverbanks ComPOOst: It's Heavy Doodie* by Riverbanks Zoo: https://www.youtube.com/watch?v=YnDZ9yAcpnI.

EXPLORATION:

- Teacher will facilitate the setup of the compost bin. Students will begin to fill the compost bin using the following steps:
  –Gather lots of food scraps, grass clippings, and falling leaves.
  –Create layers using the soil and scraps, making sure to alternate between the soil and scraps to create layers.
  –Make sure the bin has air. Stir the compost every few days with a long stick.
  –The compost bin should stay moist, but not wet. You can sprinkle the bin with water if necessary.
  –Cover the bin with a lid or a layer of soil to avoid fruit flies.

EXPLANATION:

- Compost helps make healthy soil by decomposing the decaying scraps.
- When composting, dairy, meat, and citrus should be avoided, because as they decompose, they give off odors that may attract unwanted pests.
- Compost provides nutrients for plants.
- Compost helps soil to retain water.
- Have students observe and explain what is happening in the compost bin.
- Vocabulary: *soil*, *decompose*, *compost*, *decay*, *recycle*, *garbage*, *scraps*.

ELABORATION:

- Teacher will take students out to the compost bin. Teacher will use inquiry based questions to conduct a class discussion in order for students to investigate the properties of compost with the following questions:
  –What would happen to the scraps if we did not recycle them?
  –Describe what is happening in the compost pile.
  –How is the compost beneficial for our garden?
  –Why can t we use all items for the compost bin?
- Students will complete *Creating Compost* student investigation guide.

EVALUATION:

- Students will turn in the *Creating Compost* student investigation guide.
- Teacher will observe reactions and interactions with other students and the compost.

CREATING COMPOST

Name

Draw a picture of four different items that you can use to create compost.

Compost Item #1	Compost Item #2
Compost Item #3	Compost Item #4

Circle the items that you can compost.

Plastic Bottle Leaf Fork Apple Chicken Carrots

FIRST GRADE LESSON PLAN

Watch Those Worms Work

SUBJECT / GRADE LEVEL: First Grade - Plants and their Environment.

ACADEMIC STANDARDS AND PERFORMANCE INDICATORS:

ELA:

RL.MC.5.1 Ask and answer who, what, when, where, why, and how questions about a text; use key details to make inferences and draw conclusions in texts heard or read.

RL.MC. 6.1 Describe the relationship between the illustrations and the characters, setting or events.

RL.MC.7.1 Retell text, including beginning, middle, and end; use key details to determine the theme in a text heard or read.

RL.MC.7.2 Read or listen closely to compare and contrast familiar texts and texts in author and genre studies.

RI.MC. 5.1 Ask and answer who, what, when, where, why, and how questions about a text; refer to key details to make inferences and draw conclusions in texts heard or read.

RI.MC.6.1 Retell the central idea and identify key details to summarize a text heard, read, or viewed.

RI.MC.9.1 Ask and answer questions about known and unknown words in a text.

RI.LCS. 8.1 Identify words, phrases, illustrations, and photographs used to provide information.

RI.LCS. 8.2 Use front cover, title page, illustrations/ photographs, fonts, glossary, and table of contents to locate and describe key facts or information; describe the relationship between these features and the text.

W.MCC.1.1 Explore print and multimedia sources to write opinion pieces that introduce the topic, state an opinion, give a reason for the opinion, and provide a sense of closure.

W.MCC.1.2 Plan, revise, and edit building on personal ideas and the ideas of others to strengthen writing.

W.MCC.2.1 Explore print and multimedia sources to write informative/explanatory texts that name a topic, supply facts about the topic, and provide a sense of closure.

LESSON OBJECTIVES:

TSW investigate the role of worms as decomposers that create compost for plant growth.

MATERIALS:

- Pens / paper
- Magnifying glass
- Worms
- Soil
- Containers
- Newspaper
- Large clear bin
- *Diary of a Worm* by Doreen Cronin
- *Decomposer in the Food Chain* by Alice McGenty
- *Watch Those Worms Work* student investigation guide (attached)
- *Watch Those Worms Work* teacher guide (attached)

ENGAGEMENT:

- Teacher will pose the question, "**How do worms help plants grow?**"
- Teacher will follow the engagement question by reading *Diary of a Worm* by Doreen Cronin or *Decomposer in the Food Chain* by Alice McGenty.

EXPLORATION:

- Teacher will facilitate the setup of the worm compost bin.
- Teacher will guide students in order to begin filling the compost bin with appropriate materials.

EXPLANATION:

- Teacher will explain the vocabulary while students discuss the role of worms as decomposers in the garden.
- Teacher will guide students in the construction of a worm bin.

ELABORATION:

- Students will practice their observation and illustration skills while completing their *Watch Those Worms* student investigation guide.

EVALUATION:

- Students will turn in the *Watch Those Worms* student investigation guide.
- Teacher will observe reactions and interactions with other students and the worms.

RESOURCES:

1. *Diary of a Worm* by Doreen Cronin.
2. *Decomposer in the Food Chain* by Alice McGenty.
3. *Watch Those Worms Work* handout (attached).

TEACHER GUIDE

WATCH THOSE WORMS WORK

BACKGROUND:

Students should understand that garden waste can be fed to worms. The worms create nutrient rich compost which can be used to enrich the garden soil and feed the plants.

ENGAGEMENT:

- Teacher will pose the question, "**How do worms help plants grow?**"
- Teacher will follow the engagement question by reading *Diary of a Worm* by Doreen Cronin or *Decomposer in the Food Chain* by Alice McGenty.

EXPLORATION:

- Each student will receive one worm and a cup of soil.
- **What is the worm doing?**
- **How is the worm moving?**
- **How does the worm feel?**

EXPLANATION:

- After exploring the worms, place all the worms in a clear bin lined with newspaper and soil.
- Have students place worms inside the bin.
- Every few days "feed" the worms plant scraps, such as banana peels, broccoli, or coffee grounds. ***DO NOT ADD MEAT, DAIRY, OR CITRUS.
- At each feeding, have students explore the soil through observation and touch.
- The worms will start to eat and break down, or decompose, the food scraps and decaying plants.
- After the worm digests the food, the byproduct becomes part of the soil.
- The soil becomes very fertile and full of nutrients for plants.
- Worms burrow in the ground. When they do this they make paths for water and air to get to the plant roots.
- **How do worms contribute to plant growth?**
- **Why are worms considered to be decomposers?**
- **Why do the plants need the nutrients made by the worms?**

- **How much do you think a worm can decompose?** One pound of worms can eat about half of a pound of food scraps in one day.
- Vocabulary: *byproduct*, *soil*, *scraps*, *decay*, *decomposer*, *digest*, *nutrients*, *fertilize*.

ELABORATION:

- Use the worm compost to grow a plant.
- Compare the growth of this plant to a plant grown in non-fertilized soil.
- Before each feeding, predict what you will see:
  –How will the soil look?
  –How much scraps will the worms have decomposed?
- Students will complete *Watch Those Worms Work* student investigation guide.

EVALUATION:

- Students will turn in the *Watch Those Worms Work* student investigation guide.
- Teacher will observe reactions and interactions with other students and the worms.

WATCH THOSE WORMS WORK

Name

Draw a picture of the worm bin. Describe what is happening inside of the worm bin.

FIRST GRADE LESSON PLAN

Erosion

SUBJECT / GRADE LEVEL: First Grade - Earth's Natural Resources.

ACADEMIC STANDARDS AND PERFORMANCE INDICATORS:

ELA:

W.MCC.1.1 Explore print and multimedia sources to write opinion pieces that introduce the topic, state an opinion, give a reason for the opinion, and provide a sense of closure.

W.MCC.1.2 Plan, revise, and edit building on personal ideas and the ideas of others to strengthen writing.

W.MCC.2.1 Explore print and multimedia sources to write informative/explanatory texts that name a topic, supply facts about the topic, and provide a sense of closure.

LESSON OBJECTIVES:

TSW conduct an experiment to determine what happens to soil when wind or water interacts with it.

MATERIALS:

- 4 waterproof trays
- Sod
- Loose soil
- Watering can
- Water
- Garden hose
- A small fan
- *Erosion* student illustration exercise (attached)
- *Erosion* teacher guide (attached)

ENGAGEMENT:

- Teacher will set a tray of soil in front of students.
- Teacher will hold up a fan and a hose.
- Teacher will pose the question, "**What might happen to the soil if I turned either of these hose or fan on?**"

EXPLORATION:

- Teacher will conduct an erosion experiment with the class.

EXPLANATION:

- Teacher will explain the vocabulary while students discuss the negative effects of erosion on the garden.

ELABORATION:

- Teacher will take students on an erosion walk of the school campus and students will look for examples of erosion.
- Students will practice their observation and illustration skills while completing their *Erosion* student illustration exercise.

EVALUATION:

- Students will turn in the *Erosion* student illustration exercise.

RESOURCES:

1. *Erosion* handout (attached).

TEACHER GUIDE
EROSION

BACKGROUND:

Students should understand the erosion removes soil at an accelerated rate. Erosion can negatively affect plant growth.

ENGAGEMENT:

- Teacher will set a tray of soil in front of students. Teacher will hold up a fan and a hose.
- Teacher will pose the question, "**What might happen to the soil if I turned either of these hose or fan on?**"

EXPLORATION:

- Teacher will conduct an erosion experiment with the class using the following procedures:
  –Fill two trays: one with soil, and one with sod.
  –Hold each tray over one of the empty trays.
  –Predict what will happen when the wind hits the soil. Have students simulate wind by taking turns blowing across the top of each tray.
  –Observe the empty trays.
  –Using the same trays, lean the sod and soil filled trays on an angle against a wall, with the base of each resting inside one of the empty trays, to catch residual water and soil.
  –Pour an equal amount of water on each of the trays for 5 seconds.
  –Observe the empty trays.

EXPLANATION:

- *Erosion* happens when wind or water breaks down soil or rocks on the ground.
- There are different types of erosion: water and wind erosion are two common types.
- **Which tray lost the most soil?**
- **What happens to soil when wind blows on it? When it gets rained on?**
- **How does erosion affect our lives? Our garden?**
- **How can we prevent erosion?**
- Vocabulary: *erosion*, *wind*, *water*, *runoff*.

ELABORATION:

- Teacher will take students on an erosion walk of the school campus. Students will look for examples of erosion.
- As a class, have a discussion about the negative effects of erosion and create a plan for preventing erosion in the garden.
- Students will complete *Erosion* student investigation guide.

EVALUATION:

- Students will turn in the *Erosion* student investigation guide.

EROSION

Name

____

Draw a picture of wind and water erosion.

Water Erosion

Wind Erosion

FIRST GRADE LESSON PLAN
How Long is Your Hand?

SUBJECT / GRADE LEVEL: First Grade - Math.

ACADEMIC STANDARDS AND PERFORMANCE INDICATORS:

ELA:

W.MCC.1.2 Plan, revise, and edit building on personal ideas and the ideas of others to strengthen writing.

W.MCC.2.1 Explore print and multimedia sources to write informative/explanatory texts that name a topic, supply facts about the topic, and provide a sense of closure.

LESSON OBJECTIVES:

TSW use their hands and rulers to estimate and measure the length of objects in the garden.

MATERIALS:

- Pencils
- *How Long is Your Hand?* student investigation guide (attached)
- *How Long is Your Hand?* teacher guide (attached)

ENGAGEMENT:

- Teacher will pose the question, "**What do you use to measure the length of an object?**"

EXPLORATION:

- Students should measure the width of their hand using the ruler at the top of the *How Long is Your Hand?* student investigation guide.
- Teacher will send students on a scavenger hunt to find two objects that are approximately the same width as their hand.

EXPLANATION:

- Teacher will explain the vocabulary while students investigate their hands as a unit of measurement.

ELABORATION:

- Students will complete the *How Long is Your Hand?* student investigation guide.
- Teacher will lead the class in the creation of a list of items in the garden that might need to be measured on a daily basis.

EVALUATION:

- Students will turn in the *How Long is Your Hand?* student investigation guide.
- Teacher will observe reactions and interactions with other students during the measurement exercise.

RESOURCES:

1. *How Long is Your Hand* handout (attached).

TEACHER GUIDE
HOW LONG IS YOUR HAND?

BACKGROUND:

Students should understand that their hand can be used as a point of reference for measuring objects when there is no ruler to be used.

ENGAGEMENT:

- Teacher will pose the question, "**What do you use to measure the length of an object?**"

EXPLORATION:

- Teacher will show students how to make a hand width, by spreading his or her hand on the surface of a paper and marking the outer tips of the thumbs and pinky. Using a ruler, connect the two points with a straight line and label the length "My Hand Width".
- Students should measure the width of their hand using the ruler at the top of the *How Long is Your Hand?* student investigation guide.
- Send students on a scavenger hunt to find two objects that are approximately the same width as their hand.
- Students will record the objects on the *How Long is Your Hand?* student investigation guide.
- Re-gather the group and have the children share the objects. Ask them how their hand width measurements could help them in the garden.

EXPLANATION:

- Gardeners often use hand widths, paces, and other non-traditional measuring devices to measure plants.
- Have children demonstrate their hand width with a partner by holding up their hands. Do any children have the same hand span?
- The teacher should model how to measure an object using the hand width. Ask a student to measure a second object that is longer than one hand width. Remind the student to have a fully extended hand width while measuring. As a group, determine how to count partial widths, such as "about half" a width or "a bit more" than one width.
- Gardeners still use hand widths and foot lengths as a quick way to estimate lengths of things in the garden. These include: how much room to leave between plants, stepping stones, and rows.

- Vocabulary: *hand span*, *measurement*, *standard units of measure*, *nonstandard units of measure*.

ELABORATION:

- Double Hand Width Hunt: Have the students find two things in the garden that are approximately twice the width of their hand.
- Students will complete *How Long is Your Hand?* student investigation guide.
- As a class, create a list of items in the garden you might need to measure on a daily basis.

EVALUATION:

- Students will turn in the *How Long is Your Hand?* student investigation guide.
- Teacher will observe reactions and interactions with other students during the measurement exercise.

HOW LONG IS YOUR HAND?

Name ____

Use the ruler to measure the width of your hand.

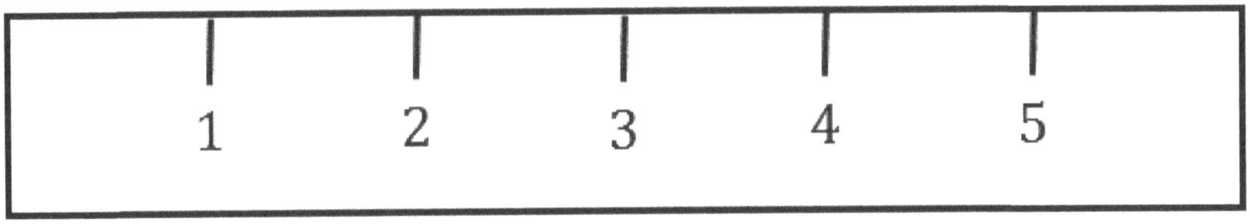

1. *How wide is your hand?* ____________________

2. *List the two objects that were the same width as your hand.*

   ____

   ____

3. *List the two objects that were twice as wide as your hand.*

   ____

   ____

FIRST GRADE LESSON PLAN

Super Sunlight

SUBJECT / GRADE LEVEL: First Grade - Exploring the Sun and Moon.

ACADEMIC STANDARDS AND PERFORMANCE INDICATORS:

ELA:

RI.MC. 5.1 Ask and answer who, what, when, where, why, and how questions about a text; refer to key details to make inferences and draw conclusions in texts heard or read.

RI.MC.6.1 Retell the central idea and identify key details to summarize a text heard, read, or viewed.

RI.MC.9.1 Ask and answer questions about known and unknown words in a text.

RL.MC.5.1 Ask and answer who, what, when, where, why, and how questions about a text; use key details to make inferences and draw conclusions in texts heard or read.

RL.MC.6.1 Describe the relationship between the illustrations and the characters, setting or events.

RL.LCS.11.2 Distinguish who is telling the story at various points in a text, the narrator or characters.

W.MCC.1.1 Explore print and multimedia sources to write opinion pieces that introduce the topic, state an opinion, give a reason for the opinion, and provide a sense of closure.

W.MCC.1.2 Plan, revise, and edit building on personal ideas and the ideas of others to strengthen writing.

W.MCC.2.1 Explore print and multimedia sources to write informative/explanatory texts that name a topic, supply facts about the topic, and provide a sense of closure.

LESSON OBJECTIVES:

TSW investigate the role of sunlight in the survival of plants.

MATERIALS:

- Plants
- A dark area such as a closet
- A bright area with natural light such as a window
- *Brown Cow, Green Grass, Yellow Mellow Sun* by Ellen Jackson
- *Super Sunlight* student investigation guide (attached)
- *Super Sunlight* teacher guide (attached)

ENGAGEMENT:

- Teacher will pose the question, "**Can plants live without sunlight?**"
- Teacher will follow the engagement question by reading *Brown Cow, Green Grass, Yellow Mellow Sun* by Ellen Jackson.

EXPLORATION:

- Teacher will facilitate the setup of a sunlight experiment involving two identical plants.

EXPLANATION:

- Teacher will explain the vocabulary while students discuss the importance of sunlight to plants.

ELABORATION:

- Students will practice their observation skills while completing their *Super Sunlight* student investigation guide over the ten day experiment.

EVALUATION:

- Students will turn in the *Super Sunlight* student investigation guide.
- Teacher will observe reactions and interactions with other students and the plants.

RESOURCES:

1. *Brown Cow, Green Grass, Yellow Mellow Sun* by Ellen Jackson.
2. *Super Sunlight* handout (attached).

TEACHER GUIDE
SUPER SUNLIGHT

BACKGROUND:

Students should understand that plants need light to create energy through photosynthesis.

ENGAGEMENT:

- Teacher will pose the question, "**Can plants live without sunlight?**"
- Teacher will follow the engagement question by reading *Brown Cow, Green Grass, Yellow Mellow Sun* by Ellen Jackson.

EXPLORATION:

- Teacher will facilitate the setup of a sunlight experiment involving two identical plants.
- One plant will be placed inside a dark area, such as a closet, while the other plant will be placed in a well-lit area with natural light, such as the window.
- Students will water the plants daily.
- Students will measure and record the height of the plants and observe the health of the plants.
- After ten days students will determine which was best for the plant: living in the light or living in the dark?

EXPLANATION:

- Leaves collect sunlight and create energy through a process called *photosynthesis*.
- The more sunlight the plant gets, the more energy the plant produces.
- Plants need sunlight to grow.
- **How do different amounts of light affect plant growth?**
- **What do farmers do in the winter months when there is less sunlight?**
- **Why is photosynthesis important?**
- Vocabulary: *photosynthesis*, *energy*, *sunlight*.

ELABORATION:

- Teacher could set up and additional experiment to test the movement of plants towards light. For this experiment, plants would be placed with the light source on one side. The class would observe how the plant turns towards the light over a period of time.
- Students will complete the *Super Sunlight* student investigation guide.

EVALUATION:

- Students will turn in the *Super Sunlight* student investigation guide.
- Teacher will observe reactions and interactions with other students and the plants.

SUPER SUNLIGHT

Name

____

Write the measurement of each plant in the correct box.

Grown in Light

Day 1	Day 2	Day 3	Day 4	Day 5	Day 6	Day 7	Day 8	Day 9	Day 10

Grown in Darkness

Day 1	Day 2	Day 3	Day 4	Day 5	Day 6	Day 7	Day 8	Day 9	Day 10

Which plants were healthiest at the end of ten days?

____

____

____

FIRST GRADE LESSON PLAN
Lovely Lunar Gardening

SUBJECT / GRADE LEVEL: First Grade - Exploring the Sun and the Moon.

ACADEMIC STANDARDS AND PERFORMANCE INDICATORS:

ELA:

RI.LCS. 8.1 Identify words, phrases, illustrations, and photographs used to provide information.

RI.LCS.8.2 Use front cover, title page, illustrations/ photographs, fonts, glossary, and table of contents to locate and describe key facts or information; describe the relationship between these features and the text.

RL.MC.5.1 Ask and answer who, what, when, where, why, and how questions about a text; use key details to make inferences and draw conclusions in texts heard or read.

RL.MC.7.1 Retell a familiar text; identify beginning, middle, and end in a text heard or read.

RL.MC.7.2 Read or listen closely to compare and contrast familiar texts and texts in author and genre studies.

W.MCC.1.1 Explore print and multimedia sources to write opinion pieces that introduce the topic, state an opinion, give a reason for the opinion, and provide a sense of closure.

W.MCC.1.2 Plan, revise, and edit building on personal ideas and the ideas of others to strengthen writing.

W.MCC.2.1 Explore print and multimedia sources to write informative/explanatory texts that name a topic, supply facts about the topic, and provide a sense of closure.

LESSON OBJECTIVES:

TSW use the moon phases to guide planting of food crops.

MATERIALS:

- Seed packets of root and leafy food crops
- Oreo cookies (approximately 2 packs), or enough for 4 cookies per child
- Calendar with moon phases
- Farmer's almanac (optional_
- Popsicle sticks
- *The Moon Book* by Gail Gibbons
- *Lovely Lunar Gardening* student investigation guide (attached)
- *Lovely lunar Gardening* teacher guide (attached)

ENGAGEMENT:

- Teacher will pose the question, "**What does the moon have to do with plants?**"
- Teacher will follow the engagement question by reading *The Moon Book* by Gail Gibbons to the class.
- Teacher will show the following video to the class: *Lunar Cycle, Why The Moon Change Shapes, 8 Phases Of The Moon, Learning Videos For Children* By Kids Educ: https://www.youtube.com/watch?v=NCweccNOaqo.

EXPLORATION:

- Teacher will facilitate a moon phase activity using cookies.
- Teacher will give examples of food crops that crop well with the phases of the moon.

EXPLANATION:

- Teacher will explain the vocabulary while students discuss the phases of the moon.

ELABORATION:

- Students will illustrate the phases of the moon while completing the *Lovely Lunar Gardening* student investigation guide.

EVALUATION:

- Students will turn in the *Lovely Lunar Gardening* student investigation guide.
- Teacher will observe reactions and interactions with other students.

RESOURCES:

1. *The Moon Book* by Gail Gibbons.
2. *Lunar Cycle, Why The Moon Change Shapes, 8 Phases Of The Moon, Learning Videos For Children* by Kids Educ: https://www.youtube.com/watch?v=NCweccNOaqo.
3. *Lovely Lunar Gardening* handout (attached).

TEACHER GUIDE
LOVELY LUNAR GARDENING

BACKGROUND:

Students should understand that the phases of the moon can affect the planting of food crops.

ENGAGEMENT:

- Teacher will pose the question, "**What does the moon have to do with plants?**"
- Teacher will follow the engagement question by reading *The Moon Book* by Gail Gibbons to the class.
- Teacher will show the following video to the class: *Lunar Cycle, Why The Moon Change Shapes, 8 Phases Of The Moon, Learning Videos For Children* by Kids Educ. https://www.youtube.com/watch?v=NCweccNOaqo.

EXPLORATION:

- Teacher will facilitate a moon phase activity using cookies. Teacher will give examples of food crops that crop well with the phases of the moon.
- **DAY 1:** Using Oreo cookies, make a model of the moon's phases. ***Be sure to demonstrate how to correctly twist the Oreo cookies to properly create the phases. Check in advance for potential food allergy issues.
- Distribute Popsicle sticks for students to manipulate the icing if needed.
- Identify the waning and waxing cycle of the moon.
- Discuss the shapes of the moon in halves and quarters.
- **DAY 2:** Using the moon phases as a guide, plan out which crops to plant in the garden by sorting seeds according to which will grow during the waning or waxing moon.
- Plant the seeds in the garden according to which phase the moon is in.
- Examples of Seeds for Waning: radishes, carrots, beets, onions.
- Examples of Seeds for Waxing: lettuce, kale, spinach, broccoli, cauliflower.

EXPLANATION:

- The moon affects the moisture of the soil.
- Moonlight also affects plants because it helps the seedlings to grow.
- During the waxing moon the light increases, while during the waning moon the light decreases.

- Each month the moon goes through a cycle. It takes 29 days for the moon to complete one cycle.
- **Why does the moon change shape?**
- **Why would a farmer follow a lunar cycle to plant crops?**
- Vocabulary: *waning*, *waxing*, *lunar*, *phase/phases*, *moon*, *increase/decrease*, *moisture*, *half*, *quarter*.

ELABORATION:

- As a class or individually, chart the moon phases for an entire month on a blank calendar.
- Students will illustrate the phases of the moon while completing the *Lovely Lunar Gardening* student investigation guide.

EVALUATION:

- Students will turn in the *Lovely Lunar Gardening* student investigation guide.
- Teacher will observe reactions and interactions with other students.

LOVELY LUNAR GARDENING

Name

____

Draw a picture of each phase of the moon and label each phase correctly.

FIRST GRADE LESSON PLAN
What Plant Part Am I Eating?

SUBJECT / GRADE LEVEL: First Grade - Plants and their Environment.

ACADEMIC STANDARDS AND PERFORMANCE INDICATORS:

ELA:

RL.MC.5.1 Ask and answer who, what, when, where, why, and how questions about a text; use key details to make inferences and draw conclusions in texts heard or read.

RL.MC.6.1 Describe the relationship between the illustrations and the characters, setting or events.

RL.MC.7.1 Retell a familiar text; identify beginning, middle, and end in a text heard or read.

RL.MC.7.2 Read or listen closely to compare and contrast familiar texts and texts in author and genre studies.

RL.MC.8.1 Read or listen closely to: a. describe characters actions, and feelings; b. compare and contrast characters experiences to those of the reader; c. describe setting; d. identify the plot including problem and solution; and e. describe cause and effect relationships.

W.MCC.1.1 Explore print and multimedia sources to write opinion pieces that introduce the topic, state an opinion, give a reason for the opinion, and provide a sense of closure.

W.MCC.1.2 Plan, revise, and edit building on personal ideas and the ideas of others to strengthen writing.

W.MCC.2.1 Explore print and multimedia sources to write informative/explanatory texts that name a topic, supply facts about the topic, and provide a sense of closure.

MATERIALS:

- Salad spinner
- Seasonal vegetables representing different parts of the plant. Examples: Carrots root, Celery stem, Lettuce leaf, Broccoli flower
- Small cups and forks for tasting
- Knife for cutting/chopping
- Cutting board
- *Muncha Muncha Mucha* by Candice Fleming
- *What Plant Part Am I Eating?* student investigation guide (attached)
- *What Plant Part Am I Eating?* teacher guide (attached)

LESSON OBJECTIVES:

TSW identify different plant parts and explain their function. TSW eat vegetables from the garden while identifying the individual parts of the plants that they consume.

ENGAGEMENT:

- Teacher will pose the question, "**What part of the plant are you eating when you eat a carrot?**"
- Teacher will follow the engagement question by reading *Muncha Muncha Mucha* by Candice Fleming to the class.

EXPLORATION:

- Teacher will facilitate the harvesting of vegetables from the garden based on the part of the plant each vegetable represents.

EXPLANATION:

- Teacher will explain the vocabulary while students discuss the consumable parts of a vegetable plant.

ELABORATION:

- Students will practice their observation and illustration skills while completing their *What Plant Part Am I Eating?* student investigation guide.

EVALUATION:

- Students will turn in the *What Plant Part Am I Eating?* student investigation guide.
- Teacher will observe reactions and interactions with other students and the vegetables.

RESOURCES:

1. *Muncha Muncha Mucha* by Candice Fleming.
2. *What Plant Part Am I Eating* handout (attached).

TEACHER GUIDE

WHAT PLANT PART AM I EATING?

BACKGROUND:

Students should understand that each time they eat a vegetable, they are consuming an important part of a plant.

ENGAGEMENT:

- Teacher will pose the question, "**What part of the plant are you eating when you eat a carrot?**"
- Teacher will follow the engagement question by reading *Muncha Muncha Mucha* by Candice Fleming to the class.

EXPLORATION:

- As a class, make a list of which vegetables can be found in the garden. Be sure to list the number of each vegetable needed so that nothing is wasted.
- Go to the garden and harvest the vegetables.
- Once the vegetables have been harvested, discuss the different parts of the plants roots, stem, leaves, flower, fruit.
- Wash the plants.
- Sort the vegetables by which part is edible. For example, carrots are roots, lettuce is a leaf, and celery is a stem.
- The teacher should assist with cutting the vegetables.

EXPLANATION:

- Each part of the plant works together to help the plant grow.
- Roots help the plant stay secure in the ground and transport water and dissolved nutrients from the soil to the plant stem.
- The stem keeps the plant up right, holds leaves in the best position to collect sunlight and transports the dissolved nutrients through the plant.
- The leaves collect sunlight to make energy.
- The flower produces the seeds when pollinated.
- Discuss why people eat some parts of a plant but not others.
- Discuss the reasons for washing the plants before eating them.

- Vocabulary: *stem, leaf, root, vegetable, salad, edible, harvest, transport, nutrient, energy, fruit, flower, salad spinner.*

ELABORATION:

- Students will complete *What Plant Parts Am I Eating?* student investigation guide.
- If time allows, create a class cookbook with recipes involving vegetables from the garden.

EVALUATION:

- Students will turn in the *What Plant Parts Am I Eating?* student investigation guide.
- Teacher will observe reactions and interactions with other students and the vegetables.

WHAT PLANT PART AM I EATING?

Name ____

Draw your favorite vegetable plant and label the parts.

Circle the parts of a plant.

Root Stem Shovel Seed Leaf
Bug

FIRST GRADE LESSON PLAN
Bean Tent Building

SUBJECT / GRADE LEVEL: First Grade - Earth's Natural Resources.

ACADEMIC STANDARDS AND PERFORMANCE INDICATORS:

ELA:

W.MCC.1.1 Explore print and multimedia sources to write opinion pieces that introduce the topic, state an opinion, give a reason for the opinion, and provide a sense of closure.

W.MCC.1.2 Plan, revise, and edit building on personal ideas and the ideas of others to strengthen writing.

W.MCC.2.1 Explore print and multimedia sources to write informative/explanatory texts that name a topic, supply facts about the topic, and provide a sense of closure.

LESSON OBJECTIVES:

TSW construct a tent bean tent using pole beans and sticks for structural support.

MATERIALS:

- Trowel or shovel
- Twine
- Pole-bean seeds
- 6 wooden sticks/poles, each one about seven feet tall. Try to find wood with rough edges.
- *Bean Tent Building* student investigation guide (attached)
- *Bean Tent Building* teacher guide (attached)

ENGAGEMENT:

- Teacher will pose the question, "**How would a tent or teepee help to grow plants in the garden?**"
- Teacher will follow the engagement question by showing the following video to the class: *How to Grow a Green Bean Teepee* by SCFarmersMarket: https://www.youtube.com/watch?v=5ausKEls_-U.

EXPLORATION:

- Teacher will facilitate the construction of the tent and planting of the beans.

EXPLANATION:

- Teacher will explain the vocabulary while students make predictions on the growth of the bean plants.

ELABORATION:

- Students will practice their observation and illustration skills while completing their *Bean Tent Building* student investigation guide.

EVALUATION:

- Students will turn in the *Bean Tent Building* student investigation guide.
- Teacher will observe reactions and interactions with other students while construction the bean tent.

RESOURCES:

- *How to Grow a Green Bean Teepee* by SCFarmersMarket: https://www.youtube.com/watch?v=5ausKEls_-U.
- *Bean Tent Building* handout (attached).

TEACHER GUIDE
BEAN TENT BUILDING

BACKGROUND:

Students should understand that vining plants, such as pole beans, need structural support in which they can climb while receiving adequate sunlight.

ENGAGEMENT:

- Teacher will pose the question, "**How would a tent or teepee help to grow plants in the garden?**"
- Teacher will follow the engagement question by showing the following video to the class: *How to Grow a Green Bean Teepee* by SCFarmersMarket: https://www.youtube.com/watch?v=5ausKEls_-U.

EXPLORATION:

- In the garden, clear a round space about six feet in diameter. Make sure the space receives direct sunlight.
- Enrich the soil with compost.
- With assistance, place the wood poles around the inner edges of the circle. Make sure there is enough space between two of the poles for an entry way.
- Place the remaining poles evenly around the circle.
- Sink the poles into the ground by holding onto them and pushing them down. Make sure each pole goes about twelve inches into the ground.
- Tie the tops of the poles together with twine about twelve inches from the top.
- Once poles are up, wrap and twist twine around each pole to provide a trellis in which the beans will grow.
- Plant pole bean seeds along the outer edge of the circle. Do not plant the pole beans between the two poles framing the entrance.
- Cover the seeds with soil.
- Water the seeds regularly and observe what happens.

EXPLANATION:

- Predict how high the bean plants will grow.
- **Why did we have to build the bean teepee in direct sunlight?**

- **What would happen if the bean teepee was not there? Where would the beans grow?**
- Vocabulary: *teepee*, *direct sunlight*, *circular*, *diameter*, *entrance*.

ELABORATION:

- Students will complete *Bean Tent Building* student investigation guide.

EVALUATION:

- Students will turn in the *Bean Tent Building* student investigation guide.
- Teacher will observe reactions and interactions with other students while construction the bean tent.

FIRST GRADE LESSON PLAN
Pollinators Love Petals

SUBJECT / GRADE LEVEL: First Grade - Plants and their Environment.

ACADEMIC STANDARDS AND PERFORMANCE INDICATORS:

ELA:

RL.MC.5.1 Ask and answer who, what, when, where, why, and how questions about a text; use key details to make inferences and draw conclusions in texts heard or read.

RL.MC.7.1 Retell a familiar text; identify beginning, middle, and end in a text heard or read.

W.MCC.1.1 Explore print and multimedia sources to write opinion pieces that introduce the topic, state an opinion, give a reason for the opinion, and provide a sense of closure.

W.MCC.1.2 Plan, revise, and edit building on personal ideas and the ideas of others to strengthen writing.

W.MCC.2.1 Explore print and multimedia sources to write informative/explanatory texts that name a topic, supply facts about the topic, and provide a sense of closure.

LESSON OBJECTIVES:

TSW understand that the function of flower petals is to attract pollinators.

MATERIALS:

- Paper
- Pencils
- Magnifying lenses
- *Bee and Me* by Elle McGuinness
- *Oh Say Can You Seed* by Dr. Seuss
- Camera, iPad, or a tablet with camera
- Pictures of various pollinators
- *Pollinators Love Petals* student investigation guide (attached)
- *Pollinators Love Petals* teacher guide (attached)

ENGAGEMENT:

- Teacher will pose the question, "**Why are flower petals colorful?**"
- Teacher will follow the engagement question by reading *Bee and Me* by Elle McGuinness or *Oh Say Can You Seed* by Dr. Seuss to the class.

EXPLORATION:

- Teacher will take students to the garden to observe pollinators in action.
- Students will take pictures to document their findings.

EXPLANATION:

- Teacher will explain the vocabulary while students discuss the important role of pollinators.

ELABORATION:

- Students will practice their observation and illustration skills while completing their *Pollinators Love Petals* student investigation guide.

EVALUATION:

- Students will turn in the *Pollinators Love Petals* student investigation guide.
- Teacher will observe reactions and interactions with other students and the pollinators.

RESOURCES:

1. *Bee and Me* by Elle J. McGuinness.
2. *Oh Say Can You Seed* by Dr. Seuss.
3. *Pollinators Love Petals* handout (attached).

TEACHER GUIDE
POLLINATORS LOVE PETALS

BACKGROUND:

Students should understand that plants use color and scent to attract pollinators. Students should understand that pollination is required for plants to produce seeds.

ENGAGEMENT:

- Teacher will pose the question, "**Why are flower petals colorful?**"
- Teacher will follow the engagement question by reading *Bee and Me* by Elle McGuinness or *Oh Say Can You Seed* by Dr. Seuss to the class.

EXPLORATION:

- Show students pictures of pollinators, including bees, butterflies, etc. Ask the students, **can you identify this insect? Why is this insect important?**
- Observe a flower in the garden. Ask the students to note any "visitors" to the flower while it's under observation. Keep a list of visitors.
- Take pictures of insects, birds, butterflies and other pollinators as they visit the flowers.
- Review the pictures taken in the garden and discuss what the insects are doing while they visit.

EXPLANATION:

- Explain that pollinators pollinate flowers and make new seeds possible.
- Discuss which pollinators visited the flower during the observation.
- Examples of pollinators include bees, butterflies, and birds.
- Pollinators such as bats as nocturnal moths visit plants at night.
- **What do pollinators do?**
- **What would happen if there were no pollinators?**
- **What would happen if there were no plants for the pollinators?**
- **What can harm pollinators?**
- **What can we do to protect pollinators?**
- Vocabulary: pollinator, pollination, flower, nectar, investigate, attract.

ELABORATION:

- Conduct an experiment to determine which flower colors and shapes attract the most pollinators.
- Conduct an experiment to determine which insects preferred certain colors or certain types of plants.
- Create a pollinators garden that will attract an array of pollinators.
- Students will complete *Pollinators Love Petals* student investigation guide.

EVALUATION:

- Students will turn in the *Pollinators Love Petals* student investigation guide.
- Teacher will observe reactions and interactions with other students and the pollinators.

POLLINATORS LOVE PETALS

Name ____

Draw a picture of four different pollinators that visit the garden.

Pollinator #1	Pollinator #2
Pollinator #3	Pollinator #4

Circle the pollinators.

Butterfly Bat Crab Fish Bee Hummingbird

FIRST GRADE LESSON PLAN
Trusted Trees

SUBJECT / GRADE LEVEL: First Grade - Earth's Natural Resources.

ACADEMIC STANDARDS AND PERFORMANCE INDICATORS:

ELA:

RI.LCS.8.1 Identify words, phrases, illustrations, and photographs used to provide information.

RI.MC.7.1 Compare and contrast topics or ideas within a thematic or author study heard, read, or viewed.

C.MC.1.4 Participate in shared conversations with varied partners about focused grade level topics and texts in small and large groups.

W.MCC.1.1 Explore print and multimedia sources to write opinion pieces that introduce the topic, state an opinion, give a reason for the opinion, and provide a sense of closure.

W.MCC.1.2 Plan, revise, and edit building on personal ideas and the ideas of others to strengthen writing.

W.MCC.2.1 Explore print and multimedia sources to write informative/explanatory texts that name a topic, supply facts about the topic, and provide a sense of closure.

LESSON OBJECTIVES:

TSW understand the importance of trees. TSW identify the ways we depend on trees in daily life.

MATERIALS:

- Small cross section of a tree also know as a tree cookie and can be obtained from a local arborist
- *From Tree to Paper* by Pam Marshall
- *The Giving Tree* by Shel Silverstein
- *A Tree is Growing* by Doris Dorros
- *Trees* by Henry Pluckrose
- *Trusted Trees* student investigation guide (attached)
- *Trusted Trees* teacher guide (attached)

ENGAGEMENT:

- Teacher will hold up a wooden pencil and pose the question, "**Where do pencils come from?**"
- Teacher will follow the engagement question by reading *From Tree to Paper* by Pam Marshall, *The Giving Tree* by Shel Silverstein, *A Tree is Growing* by Doris Dorros, or *Trees* by Henry Pluckrose to the class.

EXPLORATION:

- Teacher will pass a tree cross section around the class.
- Students will observe the wood and make predictions about the age of the tree.

EXPLANATION:

- Teacher will explain the vocabulary while students discuss the benefits of using compost in the garden.

ELABORATION:

- Teacher will use inquiry based questions to conduct a class discussion in order for students to investigate the importance of trees.
- Students will practice their observation and illustration skills while completing their T*rusted Trees* student investigation guide.

EVALUATION:

- Students will turn in the *Trusted Trees* student investigation guide.
- Teacher will observe reactions and interactions with other students and the tree cross section.

RESOURCES:

1. *From Tree to Paper* by Pam Marshall.
2. *The Giving Tree* by Shel Silverstein.
3. *A Tree is Growing* by Doris Dorros.
4. *Trees* by Henry Pluckrose.
5. *Trusted Trees* handout (attached).

TEACHER GUIDE
TRUSTED TREES

BACKGROUND:

Students should understand that trees are important natural resources which provide food and shelter to animals and humans.

ENGAGEMENT:

- Teacher will hold up a wooden pencil and pose the question, "**Where do pencils come from?**"
- Teacher will follow the engagement question by reading *From Tree to Paper* by Pam Marshall, *The Giving Tree* by Shel Silverstein, *A Tree is Growing* by Doris Dorros, or *Trees* by Henry Pluckrose to the class.

EXPLORATION:

- Pass the tree cross section around to all the students in the class. Have students look at and feel the texture of the wood.
- Estimate the age of the tree was when it was cut then count the rings on the cross section to see if estimations are correct.

EXPLANATION:

- Trees are a renewable resource.
- Trees can be replanted through natural or human processes.
- Paper, wood, and food are three products that we get from trees.
- Different trees are used for different products. Some provide people with fruits and nuts, such as apples and pecans.
- Trees begin as seeds and then turn into saplings, which are little trees.
- Trees also provide shade and help produce the oxygen that we breathe.
- Trees provide habitats for many animals.
- A tree cookie is a cross section of a tree.
- **What would happen if we did not have trees?**
- **What can we do to protect trees?**
- **What can we do to recycle more materials so that we use less trees?**
- **Why do we need to continue to plant trees?**
- **What products come from trees that we use every day?**

- **Why are trees important to us?**
- Vocabulary: *tree*, *tree cookie*, *renewable*, *resource*, *replenish*, *natural*, *texture*, *wood*, *conserve*, *sapling*.

ELABORATION:

- List items in the classroom that come from trees, such as pencils and paper.
- Research other products that are made from trees.
- Have students sort pictures of different products into three different categories: paper, wood, and food.
- Students will complete *Trusted Trees* student investigation guide.

EVALUATION:

- Students will turn in the *Trusted Trees* student investigation guide.
- Teacher will observe reactions and interactions with other students and the tree cross sections.

TRUSTED TREES

Name

____

Draw a picture of the cross section of the tree trunk including the rings.

1. *How old was this tree?*

   ____

2. *List three classroom objects that came from a tree.*

   ____

   ____

   ____

LESSONS FOR SECOND GRADE

SECOND GRADE LESSON PLAN
Investigative Gardeners

SUBJECT / GRADE LEVEL: Second Grade - Animals and their Environments.

ACADEMIC STANDARDS AND PERFORMANCE INDICATORS:

ELA:

W.MCC.2.2 Plan, revise, and edit, focusing on a topic while building on personal ideas and the ideas of others to strengthen writing.

W.L.4.8 Produce, expand, and rearrange complete simple and compound sentences.

C.MC.1.4 Participate in shared conversations with varied partners about focused grade level topics and texts in small and large groups.

RL.MC.5.1 Ask and answer literal and inferential questions to demonstrate understanding of a text; use specific details to make inferences and draw conclusions in texts heard or read.

RL.LCS.10.1 Use context to determine the meaning of words and phrases.

RI.LCS.9.4 Use print and multimedia resources to determine or clarify the precise meaning of words and phrases.

Science:

2.L.5B.1 Obtain and communicate information to describe and compare how animals interact with other animals and plants in the environment.

2.L.5B.3 Analyze and interpret data from observations to describe how animals respond to changes in their environment such as changes in food availability, water, or air.

Math:

2.MDA.1 Select and use appropriate tools e.g., rulers, yardsticks, meter sticks, measuring tapes to measure the length of an object.

2.MDA.2 Measure the same object or distance using a standard unit of one length and then a standard unit of a different length and explain verbally and in writing how and why the measurements differ.

2.MDA.3 Estimate and measure length/distance in customary units i.e., inch, foot, yard and metric units i.e., centimeter, meter.

2.MDA.4 Measure to determine how much longer one object is than another, using standard length units.

2.MDA.10 Draw conclusions from T-Charts, object graphs, picture graphs, and bar graphs.

LESSON OBJECTIVES:

TSW investigate a garden study plot to observe plant and animal interactions over time and see how the environment changes due to these interactions.

MATERIALS:

- One or more observation study plots within the garden
- Wood stakes and string to set up observation study plot, if necessary
- Camera optional
- Outdoor thermometer located in the garden area.
- *Investigative Gardeners* student investigation guide (attached)
- *Investigative Gardeners* teacher guide (attached)

ENGAGEMENT:

- Teacher will pose the question, "**What does the word observe mean?**"
- Ask students what they may want to learn about the plants, animals and insects that come into the garden.

EXPLORATION:

- Teacher will take students to the garden observation plot.
- Teacher will explain to the students that they are going to be investigators involved in a garden observation study over the next few months.

EXPLANATION:

- Teacher will use inquiry based questions to conduct a class discussion in order for students to discuss the changes that take place throughout the school year in the garden.

ELABORATION:

- Students will practice their observation and recording skills while completing their *Investigative Gardeners* student investigation guide.

EVALUATION:

- Students will turn in the *Investigative Gardeners* student investigation guide.
- Teacher will observe reactions and interactions with other students and the garden observation plot.

RESOURCES:

1. *Investigative Gardeners* handout (attached).

TEACHER GUIDE
INVESTIGATIVE GARDENERS

BACKGROUND:

Students should understand that the plants and animals within a garden change throughout the school year.

ENGAGEMENT:

- Teacher will pose the question, "**What does the word observe mean?**"
- Ask students what they may want to learn about the plants, animals and insects that come into the garden.

EXPLORATION:

- Explain to the students that they are going to be investigators involved in a garden observation study over the next few months. They will be watching the garden at different times and discovering the following:
  –How the plants grow and change.
  –What animals and insects come into the garden?
  –How the animals and insects use the plants in the study area.
- Take the students out to the garden observation plot and have them observe their surroundings.
- Students will record their weekly observations on their *Investigative Gardeners* student investigation guide. Teacher should use a new sheet each week and students should compile them in a garden journal.
- Let the students go back to the garden observation area weekly to observe how the plants, animals or insects and their environment stay the same or change over time.

EXPLANATION:

- Teacher will use inquiry based questions to conduct a class discussion in order for students to discuss the changes that take place throughout the school year in the garden.
- Ask the students to share something that they observed.
  –What animals and insects did you see using the plants?
  –What parts were they using and how did they use them?
  –How did the plants change?
  –What else did you learn during your observations?

ELABORATION:

- Teacher will discuss findings with the students.
- Animals and insects use plants in many different ways including habitat, food, water, shelter, and space.
- The animals and insects may change over time depending on the size of the plants, the weather, the season, or even the time of day.
- The animals and insects may even change the garden observation plot over time.
- Students will complete *Investigative Gardeners* student investigation guide.

EVALUATION:

- Students will turn in the *Investigative Gardeners* student investigation guide.
- Teacher will observe reactions and interactions with other students and the garden observation plot.

INVESTIGATIVE GARDENERS

Name

____

Date __________ *Time* __________ *Temperature* __________

Describe the weather in the garden today.

____

____

____

____

Describe the plants and animals observed today. Include any changes that have taken place.

____

____

____

____

SECOND GRADE LESSON PLAN

Friction and Plant Growth

SUBJECT / GRADE LEVEL: Second Grade - Exploring Pushes and Pulls.

ACADEMIC STANDARDS AND PERFORMANCE INDICATORS:

ELA:

W.MCC.2.2 Plan, revise, and edit, focusing on a topic while building on personal ideas and the ideas of others to strengthen writing.

W.L.4.8 Produce, expand, and rearrange complete simple and compound sentences.

C.MC.1.4 Participate in shared conversations with varied partners about focused grade level topics and texts in small and large groups.

RL.MC.5.1 Ask and answer literal and inferential questions to demonstrate understanding of a text; use specific details to make inferences and draw conclusions in texts heard or read.

RL.LCS.10.1 Use context to determine the meaning of words and phrases.

RI.LCS.9.4 Use print and multimedia resources to determine or clarify the precise meaning of words and phrases.

Science:

2.P.4A.4 Conduct structured investigations to answer questions about the relationship between friction and the motion of objects.

2.P.4A.5 Define problems related to the effects of friction and design possible solutions to reduce the effects on the motion of an object.

Math:

2.MDA.1 Select and use appropriate tools e.g., rulers, yardsticks, meter sticks, measuring tapes to measure the length of an object.

2.MDA.2 Measure the same object or distance using a standard unit of one length and then a standard unit of a different length and explain verbally and in writing how and why the measurements differ.

2.MDA.3 Estimate and measure length/distance in customary units i.e., inch, foot, yard and metric units i.e., centimeter, meter.

2.MDA.4 Measure to determine how much longer one object is than another, using standard length units.

2.MDA.10 Draw conclusions from T-Charts, object graphs, picture graphs, and bar graphs.

LESSON OBJECTIVES:

TSW conduct experiments to answer questions about how friction effects plant growth and how it may affect food production.

MATERIALS:

- Felt
- Smooth surface
- Coarse sand paper
- Bean seeds
- Clear plastic cups
- Rulers
- Samples of three types of soil: clay, dry / hard, healthy soil
- *Friction and Plant Growth* student investigation guide (attached)
- *Friction and Plant Growth* teacher guide (attached)

ENGAGEMENT:

- Teacher will pose the question, "**What does the word friction mean?**"
- The teacher will swipe a piece of felt across a smooth surface, such as a table, then swipe the felt across a piece of sandpaper.
- The teacher will ask students what happened in each situation: "**Why was it harder for the felt to move across the sandpaper?**" Friction is the resistance to motion.
- Have students rub their hand across their table and then across the rug.
- Ask them if they feel friction/resistance.
- "**What experience have you had with friction before?**"

EXPLORATION:

- Teacher will facilitate the setup of the friction experiment using bean seeds and different textures of soil.

EXPLANATION:

- Teacher will use inquiry based questions to conduct a class discussion in order for students to discuss the friction experiment.

ELABORATION:

- Students will practice their observation and recording skills while completing their *Friction and Plant Growth* student investigation guide.

EVALUATION:

- Students will turn in the *Friction and Plant Growth* student investigation guide.
- Teacher will observe reactions and interactions with other students and the bean plants.

RESOURCES:

1. *Friction and Plant Growth* handout (attached).

TEACHER GUIDE

FRICTION AND PLANT GROWTH

BACKGROUND:

Students should understand that forces such as friction can affect plant growth.

ENGAGEMENT:

- Teacher will pose the question, "**What does the word friction mean?**"
- The teacher will swipe a piece of felt across a smooth surface, such as a table. The teacher will then swipe the felt across a piece of sandpaper.
- The teacher will ask students what happened in each situation. **Why was it harder for the felt to move across the sandpaper?** Friction is the resistance to motion.
- Have students rub their hand across their table and then across the rug. Ask them if they feel friction/resistance.
- **What experience have you had with friction before?**

EXPLORATION:

- Divide students into groups of three. Each group will be given three plastic cups and nine bean seeds. They will label each cup accordingly: "c" for the clay soil, "d" for the dry / hard soil, and "h" for the healthy soil, and fill the cups with the appropriate soil type.
- Students will then use their rulers to measure one inch into the soil for each of the three bean seeds in each sample cup, for a total of nine holes. Plant each seed and water slightly.
- Students will make a hypothesis regarding each sample.
- Daily observations will be made for ten days and students will record their measurements on the *Friction and Plant Growth* student investigation guide.
- **Which seeds sprouted first? Which one sprouted last? Did any not sprout at all?**

EXPLANATION:

- Teacher will use inquiry based questions to conduct a class discussion in order for students to discuss the friction experiment.
- Students will share their hypothesis and discuss their findings with the group.
- **What does this experiment show you?**
- **How does your soil sample effect the growth of your bean seed?**
- **Can you see that the more dense the soil, the more increase in friction, the less growth?**

- Vocabulary: *friction, resistance, motion, clay, sprouted, experiment.*

ELABORATION:

- The teacher will discuss the three types of soil samples with the students, and elaborate on each soil s affect on seed growth.
- **Why is it important to have healthy soil?**
- **Do different plants have different soil needs?**
- Think about how important it is for farmers to pay attention to soil quality to ensure the best food production. **What are some ways that farmers can help improve their soil?**
- Students can go out into the school garden and take a handful of soil. **Does it seem healthy?**
- **Would your plants experience much friction? What could you do to improve the soil quality?**

EVALUATION:

- Students will turn in the *Friction and Plant Growth* student investigation guide.
- Teacher will observe reactions and interactions with other students and the bean plants.

FRICTION AND PLANT GROWTH

Name ____

Circle the plant that grew the tallest.

Plant C

Plant D

Plant H

Plant C

Day 1	Day 2	Day 3	Day 4	Day 5	Day 6	Day 7	Day 8	Day 9	Day 10

Plant D

Day 1	Day 2	Day 3	Day 4	Day 5	Day 6	Day 7	Day 8	Day 9	Day 10

Plant H

Day 1	Day 2	Day 3	Day 4	Day 5	Day 6	Day 7	Day 8	Day 9	Day 10

SECOND GRADE LESSON PLAN

Garden Recovery

SUBJECT / GRADE LEVEL: Second Grade - Weather.

ACADEMIC STANDARDS AND PERFORMANCE INDICATORS:

ELA:

W.MCC.2.2 Plan, revise, and edit, focusing on a topic while building on personal ideas and the ideas of others to strengthen writing.

W.L.4.8 Produce, expand, and rearrange complete simple and compound sentences.

C.MC.1.4 Participate in shared conversations with varied partners about focused grade level topics and texts in small and large groups.

RL.MC.5.1 Ask and answer literal and inferential questions to demonstrate understanding of a text; use specific details to make inferences and draw conclusions in texts heard or read.

RL.LCS.10.1 Use context to determine the meaning of words and phrases.

RI.LCS.9.4 Use print and multimedia resources to determine or clarify the precise meaning of words and phrases.

Science:

2.E.2A.1 Analyze and interpret data from observations and measurements to describe local weather conditions including temperature, wind, and forms of precipitation.

2.E.2A.2 Analyze local weather data to predict daily and seasonal patterns over time.

2.E.2A.3 Develop and use models to describe and compare the effects of wind moving air on objects.

2.E.2A.4 Obtain and communicate information about severe weather conditions to explain why certain safety precautions are necessary.

Math:

2.MDA.1 Select and use appropriate tools e.g., rulers, yardsticks, meter sticks, measuring tapes to measure the length of an object.

2.MDA.2 Measure the same object or distance using a standard unit of one length and then a standard unit of a different length and explain verbally and in writing how and why the measurements differ.

2.MDA.3 Estimate and measure length/distance in customary units i.e., inch, foot, yard and metric units i.e., centimeter, meter.

2.MDA.4 Measure to determine how much longer one object is than another, using standard length units.

2.MDA.10 Draw conclusions from T-Charts, object graphs, picture graphs, and bar graphs.

LESSON OBJECTIVES:

TSW follow storm weather safety procedures to care for themselves and the garden after a storm.

MATERIALS:

- Gloves
- Pencil and paper to record observations
- Trash container for debris
- *Franklin and the Thunderstorm* by Paulette Bourgeois
- *Garden Recovery* student investigation guide (attached)
- *Garden Recovery* teacher guide (attached)

ENGAGEMENT:

- Teacher will pose the question, "**What can happen to a garden during a storm?**"
- Teacher will follow the engagement question by reading *Franklin and the Thunderstorm* by Paulette Bourgeois to the class.

EXPLORATION:

- Led by the teacher, students will generate a list of garden preparations in the event of a storm.

EXPLANATION:

- Teacher will conduct a class discussion in order for students to discuss the damage that can happen in gardens due to storms.
- Teacher will facilitate a discussion on how to save the plants and keep people safe, too.

ELABORATION:

- Students will complete *Garden Recovery* student investigation guide.

EVALUATION:

- Students will turn in the *Garden Recovery* student investigation guide.
- Teacher will observe reactions and interactions with other students and the garden.

RESOURCES:

1. *Franklin and the Thunderstorm* by Paulette Bourgeois.
2. *Garden Recovery* handout (attached).

TEACHER GUIDE
GARDEN RECOVERY

BACKGROUND:

Students should understand that storms can cause damage to a garden. Preparation and cleanup are important parts of garden storm preparation and management.

ENGAGEMENT:

- Teacher will pose the question, "**What can happen to a garden during a storm?**"
- Teacher will follow the engagement question by reading *Franklin and the Thunderstorm* by Paulette Bourgeois to the class.

EXPLORATION:

- Led by the teacher, students will generate a list of garden preparations in the event of a storm.

EXPLANATION:

- Teacher will conduct a class discussion in order for students to discuss the damage that can happen in gardens due to storms. Teacher will facilitate a discussion on how to save the plants and keep people safe, too.
- Vocabulary: *severe*, *thunderstorm*, *perishable*, *debris*.

ELABORATION:

- Students will complete the *Garden Recovery* student investigation guide.

EVALUATION:

- Students will turn in the *Garden Recovery* student investigation guide.
- Teacher will observe reactions and interactions with other students and the garden.

GARDEN RECOVERY

Name

Draw a picture of the garden after a storm. Describe what has happened in the garden.

SECOND GRADE LESSON PLAN
Nocturnal Helpers

SUBJECT / GRADE LEVEL: Second Grade - Animals and their Environments.

ACADEMIC STANDARDS AND PERFORMANCE INDICATORS:

ELA:

W.MCC.2.2 Plan, revise, and edit, focusing on a topic while building on personal ideas and the ideas of others to strengthen writing.

W.L.4.8 Produce, expand, and rearrange complete simple and compound sentences.

C.MC.1.4 Participate in shared conversations with varied partners about focused grade level topics and texts in small and large groups.

RL.MC.5.1 Ask and answer literal and inferential questions to demonstrate understanding of a text; use specific details to make inferences and draw conclusions in texts heard or read.

RL.LCS.10.1 Use context to determine the meaning of words and phrases.

RI.LCS.9.4 Use print and multimedia resources to determine or clarify the precise meaning of words and phrases.

Science:

2.L.5B.1 Obtain and communicate information to describe and compare how animals interact with other animals and plants in the environment.

LESSON OBJECTIVES:

TSW understand that animals come out at night to eat insects that may feed on garden crops.

MATERIALS:

- Terra cotta pots
- Waterproof paint
- *Nocturnal Helpers* student investigation guide (attached)
- *Nocturnal Helpers* teacher guide (attached)

ENGAGEMENT:

- Teacher will pose the question, "**What insect eating animals come out at night?**"
- Ask students how nocturnal animals could be beneficial to plants in the garden.

EXPLORATION:

- Teacher will create student groups and each group will build a toad house for the garden.

EXPLANATION:

- Teacher will explain the vocabulary while using inquiry based questioning to conduct a class discussion about the benefits of nocturnal animals.

ELABORATION:

- Students will practice their observation and recording skills while completing their *Nocturnal Helpers* student investigation guide.

EVALUATION:

- Students will turn in the *Nocturnal Helpers* student investigation guide.
- Teacher will observe reactions and interactions with other students and the toad houses.

RESOURCES:

1. *Nocturnal Helpers* handout (attached).

TEACHER GUIDE
NOCTURNAL HELPERS

BACKGROUND:

Students should understand that nocturnal animals can eat some of the insects that feed on garden plants.

ENGAGEMENT:

- Teacher will pose the question, "**What insect eating animals come out at night?**"
- Ask students how the nocturnal animals could be beneficial to plants in the garden.

EXPLORATION:

- Create students groups. Each group will build a toad home by decorating the terra-cotta pot with paint.
- Place the pot next to the garden. Bury the edge of the container in the ground to hold it in place.
- Peek inside the toad home every few days to see if you have any visitors.
- When you find something in the home, record your observations in your science journal.

EXPLANATION:

- Toads are nocturnal, which means they only come out at night and sleep during the day.
- At night toads come out from their toad home and feast on the insects that may eat the crops.
- This helps keep the crops and toads healthy.
- **What other animals might help eat the insects in the garden?**
- **What insects do the toads eat?**
- **What would happen if there were no toads living near the garden to eat the insects?**
- Vocabulary: *insect, toad, nocturnal, feast.*

ELABORATION:

- Create a list of other animals that are nocturnal.
- Students will complete *Nocturnal Helpers* student investigation guide.

EVALUATION:

- Students will turn in the *Nocturnal Helpers* student investigation guide.
- Teacher will observe reactions and interactions with other students and the toad houses.

NOCTURAL HELPERS

Name

Draw four different nocturnal animals below.

Nocturnal Animal 1	**Nocturnal Animal 2**
Nocturnal Animal 3	**Nocturnal Animal 4**

How do nocturnal animals help plants?

SECOND GRADE LESSON PLAN

Plants and Magnets

SUBJECT / GRADE LEVEL: Second Grade - Properties of Solids and Liquids.

ACADEMIC STANDARDS AND PERFORMANCE INDICATORS:

ELA:

W.MCC.2.2 Plan, revise, and edit, focusing on a topic while building on personal ideas and the ideas of others to strengthen writing.

W.L.4.8 Produce, expand, and rearrange complete simple and compound sentences.

C.MC.1.4 Participate in shared conversations with varied partners about focused grade level topics and texts in small and large groups.

RL.MC.5.1 Ask and answer literal and inferential questions to demonstrate understanding of a text; use specific details to make inferences and draw conclusions in texts heard or read.

RL.LCS.10.1 Use context to determine the meaning of words and phrases.

RI.LCS.9.4 Use print and multimedia resources to determine or clarify the precise meaning of words and phrases.

Science:

2.P.3B.2 Analyze and interpret data from observations to compare the effects of magnets on various materials.

2.P.3B.3 Obtain and communicate information to exemplify the uses of magnets in everyday life.

Math:

2.MDA.1 Select and use appropriate tools e.g., rulers, yardsticks, meter sticks, measuring tapes to measure the length of an object.

2.MDA.2 Measure the same object or distance using a standard unit of one length and then a standard unit of a different length and explain verbally and in writing how and why the measurements differ.

2.MDA.3 Estimate and measure length/distance in customary units i.e.: inch, foot, yard and metric units i.e., centimeter, meter.

2.MDA.4 Measure to determine how much longer one object is than another, using standard length units.

2.MDA.10 Draw conclusions from T-Charts, object graphs, picture graphs, and bar graphs.

LESSON OBJECTIVES:

TSW identify the special properties of magnets. TSW use magnetized water to care for a plant.

MATERIALS:

- Large magnets variety of sizes
- Assortment of magnets and non-magnetic materials, magnetic shavings
- Plants, pots for plants
- Watering cans
- *Plants and Magnets* student investigation guide (attached)
- *Plants and Magnets* teacher guide (attached)

ENGAGEMENT:

- Teacher will pose the question, "**Do you think magnets can affect plant growth?**"

EXPLORATION:

- Students will use magnets to explore objects in containers in the room and then take the magnets to the garden to explore.

EXPLANATION:

- Teacher will divide students into groups to begin the nine week experiment involving plants grown with magnetized and non-magnetized water.

ELABORATION:

- Students will practice their recording skills while completing their *Plants and Magnets* student investigation guide.

EVALUATION:

- Students will turn in the *Plants and Magnets* student investigation guide.
- Teacher will observe reactions and interactions with other students and the magnets.

RESOURCES:

1. *Plants and Magnets* handout (attached).

TEACHER GUIDE
PLANTS AND MAGNETS

BACKGROUND:

Students should understand that magnets can have a slight effect on plant growth over time.

ENGAGEMENT:

- Teacher will pose the question, "**Do you think magnets can affect plant growth?**"

EXPLORATION:

- **Do you think magnets would be attracted to plants?**
- **Magnets are attracted to iron so do you think magnets could affect plant growth?**
- Students will use magnets to explore objects in containers in the room, and then take the magnets to the garden to explore.

EXPLANATION:

- Divide the class into groups of three to five students.
- Each group will get two plants. One plant will be magnetized, and one will not.
- Drop magnets or magnet shavings in water and let it sit a few minutes. Pour this water onto one of the plants.
- The students will take measurements of stem and leaf growth over nine weeks.
- The students will compare the growth of the magnetized plants to the non-magnetized plants and record their results.
- Vocabulary: *magnetize*, *magnetic field*.

ELABORATION:

- Students can magnetize seeds before planting, and compare the growth of non-magnetized seeds to the growth of the magnetized seeds.
- Students can research how magnetizing is affecting crop production.
- Students will complete *Plants and Magnets* student investigation guide.

EVALUATION:

- Students will turn in the *Plants and Magnets* student investigation guide.
- Teacher will observe reactions and interactions with other students and the magnets.

PLANTS AND MAGNETS

Name

____

Record the height of the plant being watered with magnetized water.

Week 1	Week 2	Week 3	Week 4	Week 5	Week 6	Week 7	Week 8	Week 9

Record the height of the plant being watered with non-magnetized water.

Week 1	Week 2	Week 3	Week 4	Week 5	Week 6	Week 7	Week 8	Week 9

Which plant grew the tallest?

____

SECOND GRADE LESSON PLAN

Frozen Flowers

SUBJECT / GRADE LEVEL: Second Grade - Properties of Solids and Liquids.

ACADEMIC STANDARDS AND PERFORMANCE INDICATORS:

ELA:

W.MCC.2.2 Plan, revise, and edit, focusing on a topic while building on personal ideas and the ideas of others to strengthen writing.

W.L.4.8 Produce, expand, and rearrange complete simple and compound sentences.

C.MC.1.4 Participate in shared conversations with varied partners about focused grade level topics and texts in small and large groups.

RL.MC.5.1 Ask and answer literal and inferential questions to demonstrate understanding of a text; use specific details to make inferences and draw conclusions in texts heard or read.

RL.LCS.10.1 Use context to determine the meaning of words and phrases.

RI.LCS.9.4 Use print and multimedia resources to determine or clarify the precise meaning of words and phrases.

Science:

2.P.3A.3 Conduct structured investigations to test how adding or removing heat can cause changes in solids and liquids.

2.P3A.4 Construct scientific arguments using evidence from investigations to support claims that some changes in solids or liquids are reversible and some are not when heat is added or removed.

Math:

2.MDA.1 Select and use appropriate tools e.g., rulers, yardsticks, meter sticks, measuring tapes to measure the length of an object.

2.MDA.2 Measure the same object or distance using a standard unit of one length and then a standard unit of a different length and explain verbally and in writing how and why the measurements differ.

2.MDA.3 Estimate and measure length/distance in customary units i.e., inch, foot, yard and metric units i.e., centimeter, meter.

2.MDA.4 Measure to determine how much longer one object is than another, using standard length units.

2.MDA.10 Draw conclusions from T-Charts, object graphs, picture graphs, and bar graphs.

LESSON OBJECTIVES:

TSW identify some endangered plants and their uses. TSW discuss the problem of disappearing plants. TSW investigate plant survival in frozen temperatures.

MATERIALS:

- 4 healthy plants
- Ice cubes
- Water
- A freezer
- Pictures of endangered plants
- *Frozen Flowers* student investigation guide (attached)
- *Frozen Flowers* teacher guide (attached)

ENGAGEMENT:

- Teacher will pose the question, "**What does it mean for a plant or animal to be endangered?**"

EXPLORATION:

- Teacher will use inquiry based questions to conduct a class discussion in order for students to discuss the importance of saving endangered species.

EXPLANATION:

- Teacher will explain that some plants can survive freezing temperatures while others cannot.
- Teacher will facilitate the setup of an experiment in which plants are frozen for two weeks.

ELABORATION:

- Students will practice their observation and recording skills while completing their *Frozen Flowers* student investigation guide.

EVALUATION:

- Students will turn in the *Frozen Flowers* student investigation guide.
- Teacher will observe reactions and interactions with other students and the frozen plants.

REOURCES:

1. *Frozen Flowers* handout (attached).

TEACHER GUIDE

FROZEN FLOWERS

BACKGROUND:

Students should understand that the plants and animals can become endangered.

ENGAGEMENT:

- Teacher will pose the question, "**What does it mean for a plant or animal to be endangered?**"

EXPLORATION:

- **How do we save the plants that are disappearing?**
- **Why do you think they are disappearing?**
- **Look at pictures of endangered plants.**
- **If an endangered animal like a panda goes extinct, is it a greater loss than if a plant goes extinct?**
- **How would a scientist decide which plants need attention?**
- **How would you suggest saving endangered plants?**
- **Are some plants more worth saving than others? Why?**
- **Is animal life dependent on plant life?**

EXPLANATION:

- **Which common species of garden plants can survive being frozen?**
- **How do plants survive being frozen? In soil, without soil?**
- **How low does the temperature have to be?**
- **Will our freezer be a good simulation for what a scientist would do?**
- Knowing how freezing temperatures affect plants can help farmers know when they need to protect their crops and when the adaptations of the plant are sufficient for it to protect itself.
- Scientists are experimenting with freezing plants to conserve them for the future.
- Take the four plants, photograph them, measure them, and label them.
- Place one in the freezer without a cover.
- Place another one in the freezer in a Ziploc bag.
- Place one more in the freezer in a paper bag.
- Place one in the refrigerator.

- Students will log weekly information on their *Frozen Flowers* student investigation guide.
- Vocabulary: *endangered*, *extinct*, *conservation*.

ELABORATION:

- **How can we keep our garden plants healthy in cold weather?**
- **Why wouldn't freezing seeds be better than freezing plants?**
- Divide students into groups. Each group chooses an endangered plant to study.
- **How would scientists determine which plants to conserve for the future?**
- **Should medical value be the first consideration?**
- Students will complete the *Frozen Flowers* student investigation guide.

EVALUATION:

- Students will turn in the *Frozen Flowers* student investigation guide.
- Teacher will observe reactions and interactions with other students and the frozen plants.

FROZEN FLOWERS

Name ____

Week One

	Date	Description of Plant
Freezer Covered		
Freezer Plastic Bag		
Freezer Paper Bag		
Refrigerator		

Week Two

	Date	Description of Plant
Freezer Covered		
Freezer Plastic Bag		
Freezer Paper Bag		
Refrigerator		

SECOND GRADE LESSON PLAN

Topsy Turvy Seeds

SUBJECT / GRADE LEVEL: Second Grade - Exploring Pushes and Pulls.

ACADEMIC STANDARDS AND PERFORMANCE INDICATORS:

ELA:

W.MCC.2.2 Plan, revise, and edit, focusing on a topic while building on personal ideas and the ideas of others to strengthen writing.

W.L.4.8 Produce, expand, and rearrange complete simple and compound sentences.

C.MC.1.4 Participate in shared conversations with varied partners about focused grade level topics and texts in small and large groups.

RL.MC.5.1 Ask and answer literal and inferential questions to demonstrate understanding of a text; use specific details to make inferences and draw conclusions in texts heard or read.

RL.LCS.10.1 Use context to determine the meaning of words and phrases.

RI.LCS.9.4 Use print and multimedia resources to determine or clarify the precise meaning of words and phrases.

Science:

2.P.4A.3 Construct explanations of the relationship between the motion of an object and the pull of gravity using observations and data collected.

Math:

2.MDA.1 Select and use appropriate tools e.g., rulers, yardsticks, meter sticks, measuring tapes to measure the length of an object.

2.MDA.2 Measure the same object or distance using a standard unit of one length and then a standard unit of a different length and explain verbally and in writing how and why the measurements differ.

2.MDA.3 Estimate and measure length/distance in customary units i.e., inch, foot, yard and metric units i.e., centimeter, meter.

2.MDA.4 Measure to determine how much longer one object is than another, using standard length units.

2.MDA.10 Draw conclusions from T-Charts, object graphs, picture graphs, and bar graphs.

LESSON OBJECTIVES:

TSW investigate the effect of gravity on newly planted seeds.

MATERIALS:

- Bean seeds
- Clear jars (recycled)
- Wet paper towels
- *Topsy Turvy Seeds* student investigation guide (attached)
- *Topsy Turvy Seeds* teacher guide (attached)

ENGAGEMENT:

- Teacher will pose the question, "**What is gravity?**"

EXPLORATION:

- Teacher will divide students into pairs and have each pair set up a geotropism experiment using bean seeds and jars.

EXPLANATION:

- Teacher will introduce the vocabulary word *geotropism*.
- Teacher will use inquiry based questions to conduct a class discussion in order for students to discuss the experiment results.

ELABORATION:

- Students will practice their observation and illustration skills while completing their *Topsy Turvy Seeds* student investigation guide.

EVALUATION:

- Students will turn in the *Topsy Turvy Seeds* student investigation guide.
- Teacher will observe reactions and interactions with other students and the bean seeds.

RESOURCES:

1. *Topsy Turvy Seeds* handout (attached).

TEACHER GUIDE
TOPSY TURVY SEEDS

BACKGROUND:

Students should understand that gravity affects not only humans but plants as well.

ENGAGEMENT:

- Teacher will pose the question, "**What is gravity?**"

EXPLORATION:

- **What do we know about how seeds and plants grow?**
- **Do you think that gravity has any effect on plant growth?**
- Divide students into pairs. Each pair will be given two glass jars, six bean seeds and two wet paper towels. Label one jar with the letter "G" and the other with "control".
- Students will line up each jar with a wet paper towel and place three beans seeds, evenly spaced, between the jar and paper towel. The paper towels should remain moist during growth time.
- Once stems are about one inch long, have students turn jar "G" on its side. Leave the control jar upright.
- Students should now make a hypothesis on what they think the stems and roots will do in their two jars.
- Students will record the daily growth of their stems and roots for two weeks. Have them use a ruler to measure in inches and centimeters. Students can draw pictures of the growth.
- Students should change position of jar "G" ever forty-eight hours, but always keep the control jar upright.

EXPLANATION:

- Teacher will introduce the vocabulary word **geotropism**. Teacher will use inquiry based questions to conduct a class discussion in order for students to discuss the experiment results.
  –What does this experiment tell you about the growth of the stem?
  –What does it tell you about the growth of roots?
  –Why is this important for plant growth?
  –Is gravity absolutely necessary for plant growth?

ELABORATION:

- Suggest that students transplant their seeds outside. Bean seedlings could be planted outside to construct a bean pole teepee.
- Research and experiment with other tropisms such as phototropism.
- Students will complete *Topsy Turvy* Seeds student investigation guide.

EVALUATION:

- Students will turn in the *Topsy Turvy Seeds* student investigation guide.
- Teacher will observe reactions and interactions with other students and the bean seeds.

TOPSY TURVY SEEDS

Name

Draw a picture of the bean seeds growing in the G jar. Describe what is happening inside of the G jar.

SECOND GRADE LESSON PLAN

Water Goes Up the Tree

SUBJECT / GRADE LEVEL: Second Grade - Properties of Solids and Liquids.

ACADEMIC STANDARDS AND PERFORMANCE INDICATORS:

ELA:

W.MCC.2.2 Plan, revise, and edit, focusing on a topic while building on personal ideas and the ideas of others to strengthen writing.

W.L.4.8 Produce, expand, and rearrange complete simple and compound sentences.

C.MC.1.4 Participate in shared conversations with varied partners about focused grade level topics and texts in small and large groups.

RL.MC.5.1 Ask and answer literal and inferential questions to demonstrate understanding of a text; use specific details to make inferences and draw conclusions in texts heard or read.

RL.LCS.10.1 Use context to determine the meaning of words and phrases.

RI.LCS.9.4 Use print and multimedia resources to determine or clarify the precise meaning of words and phrases.

Science:

2.P.3A.1 Analyze and interpret data from observations and measurements to describe the properties used to classify matter as a solid or a liquid.

2.P3A.4 Construct scientific arguments using evidence from investigations to support claims that some changes in solids or liquids are reversible and some are not when heat is added or removed.

Math:

2.MDA.1 Select and use appropriate tools e.g., rulers, yardsticks, meter sticks, measuring tapes to measure the length of an object.

2.MDA.2 Measure the same object or distance using a standard unit of one length and then a standard unit of a different length and explain verbally and in writing how and why the measurements differ.

2.MDA.3 Estimate and measure length/distance in customary units i.e., inch, foot, yard and metric units i.e., centimeter, meter.

2.MDA.4 Measure to determine how much longer one object is than another, using standard length units.

2.MDA.10 Draw conclusions from T-Charts, object graphs, picture graphs, and bar graphs.

LESSON OBJECTIVES

TSW identify the change in water as it is transported up from the tree roots to the tree top.

MATERIALS:

- A cup, half-filled with water
- Some blue or red food coloring
- A stalk of celery with some leaves on it
- *Water Goes Up the Tree* student investigation guide (attached)
- *Water Goes Up the Tree* teacher guide (attached)

ENGAGEMENT:

- Teacher will pose the question, "**How does water move up a tree?**"

EXPLORATION:

- Teacher will help students set up an experiment that will show the ability of plant stems to transport water from the roots up through the stem of the plant.

EXPLANATION:

- Teacher will conduct a class discussion in order for students to discuss the ability of plant stems to move water up from the roots in everything from small plants to large trees.

ELABORATION:

- Students will practice their observation and recording skills while completing their *Water Moves Up the Tree* student investigation guide.

EVALUATION:

- Students will turn in the *Water Moves Up the Tree* student investigation guide.
- Teacher will observe reactions and interactions with other students and the celery.

RESOURCES:

1. *Water Goes Up the Tree* handout (attached).

TEACHER GUIDE
WATER MOVES UP THE TREE

BACKGROUND:

Students should understand that once the roots absorb water then the stem must transport that water up the tree.

ENGAGEMENT:

- Teacher will pose the question, "**How does water move up a tree?**"

EXPLORATION:

- Teacher will help students set up an experiment that will show the ability of plant stems to transport water from the roots up through the stem of the plant.
- The students will mix a teaspoon of the food coloring into the water.
- Cut the celery stalk about one inch from the bottom to expose a fresh end, and stand the stalk in the water.
- Leave the celery in the water for an hour or two.
- What is happening to the leaves? The dye should gradually color the leaves.
- When the color has spread to the tips of the leaves, take the celery out of the water and cut across the stalk.
- **What do you see?** A row of tiny circles outlined in color.
- These are the ends of fine tubes that travel the length of the stalk.
- **Where did the colored water go?** The colored water traveled up those tubes xylem and phloem tubes.

EXPLANATION:

- Teacher will conduct a class discussion in order for students to discuss the ability of plant stems to move water up from the roots in everything from small plants to large trees.
- Trees have similar tubes running up their trunks.
- What makes the water climb the tress is still somewhat of a mystery, but scientists think it all depends on the special properties of water.
- The tubes are porous and very narrow. As the tubes spread out into the leaves, heat from the sun evaporates the water molecules at the top.

- In the same way that water climbs a short distance up the walls of certain substances like drinking glasses, for instance, the next molecules always hold tightly together, and when they're squished into the very narrow tubes, they grip more tightly, with enough strength to pull all the following water molecules along behind them.
- So as the molecules at the top move up, the whole chain moves up the tree. This only works, however, if the tubes are full of liquid to begin with, so trees and other plants have liquid filled tubes from their earliest days as seedlings.
- What about trees hit by lightning?
- What if there is no rain, or too much rain, like a flood?
- Vocabulary: molecules, transpiration, upward, substance, seedling.

ELABORATION:

- Students can plant the bottom of the celery plants in the garden.
- Students will complete *Water Moves Up the Tree* student investigation guide.

EVALUATION:

- Students will turn in the *Water Moves Up the Tree* student investigation guide.
- Teacher will observe reactions and interactions with other students and the celery.

WATER MOVES UP THE TREE

Name

____

What color dye did you add to the celery water?

____

How long did it take for the colored water to travel up to the leaves of the plant?

____

Describe the inside of the celery stem after it was cut in half.

____

____

____

How did the colored water travel from the bottom to the top of the stem?

____

____

____

SECOND GRADE LESSON PLAN

Weather Instruments

SUBJECT / GRADE LEVEL: Second Grade - Earth Science - Making Instruments to Analyze and Interpret Data - Weather.

ACADEMIC STANDARDS AND PERFORMANCE INDICATORS:

ELA:

W.MCC.2.2 Plan, revise, and edit, focusing on a topic while building on personal ideas and the ideas of others to strengthen writing.

W.L.4.8 Produce, expand, and rearrange complete simple and compound sentences.

C.MC.1.4 Participate in shared conversations with varied partners about focused grade level topics and texts in small and large groups.

RL.MC.5.1 Ask and answer literal and inferential questions to demonstrate understanding of a text; use specific details to make inferences and draw conclusions in texts heard or read.

RL.LCS.10.1 Use context to determine the meaning of words and phrases.

RI.LCS.9.4 Use print and multimedia resources to determine or clarify the precise meaning of words and phrases.

Science:

2.E.2A.1 Analyze and interpret data from observations and measurements to describe local weather conditions including temperature, wind, and forms of precipitation.

2.E.2A.2 Analyze local weather data to predict daily and seasonal patterns over time.

2.E.2A.3 Develop and use models to describe and compare the effects of wind moving air on objects.

2.E.2A.4 Obtain and communicate information about severe weather conditions to explain why certain safety precautions are necessary.

Math:

2.MDA.1 Select and use appropriate tools e.g., rulers, yardsticks, meter sticks, measuring tapes to measure the length of an object.

2.MDA.2 Measure the same object or distance using a standard unit of one length and then a standard unit of a different length and explain verbally and in writing how and why the measurements differ.

2.MDA.3 Estimate and measure length/distance in customary units i.e., inch, foot, yard and metric units i.e., centimeter, meter.

2.MDA.4 Measure to determine how much longer one object is than another, using standard length units.

2.MDA.10 Draw conclusions from T-Charts, object graphs, picture graphs, and bar graphs.

LESSON OBJECTIVES:

TSW make a rain gauge, anemometer, and a barometer to analyze and interpret data to describe local weather conditions including wind, temperature, and forms of precipitation.

MATERIALS:

- Rain gauge
- Anemometer
- Barometer
- *Extreme Weather Conditions* by Leanne Annette
- *Weather Instruments* student journal pages
- *Weather Instruments* teacher guide (attached)

ENGAGEMENT:

- Teacher will pose the question, "**What does a thermometer measure?**"
- Teacher will follow the engagement question by reading *Extreme Weather Conditions* by Leanne Annette to the class.

EXPLORATION:

- Teacher will take students to the garden observation plot.
- Teacher will explain to the students that they are going to be measuring weather with several different devices.

EXPLANATION:

- Teacher will show the students how to read the instruments and how the information is used to analyze and interpret weather conditions.

ELABORATION:

- Students will practice their observation and recording skills while completing their *Weather Instruments* student journal pages.

EVALUATION:

- Students will compile the *Weather Instruments* student journal pages in a binder.
- Teacher will observe reactions and interactions with other students and the weather instruments.

RESOURCES:

1. *Extreme Weather Conditions* by Leanne Annette.
2. *Weather Instruments* handout (attached).

TEACHER GUIDE

WEATHER INSTRUMENTS

BACKGROUND:

Students should understand that garden plants are affected by changes in the weather and there are several useful instruments that can help us to measure the weather.

ENGAGEMENT:

- Teacher will pose the question, "**What does a thermometer measure?**"
- Teacher will follow the engagement question by reading *Extreme Weather Conditions* by Leanne Annette to the class.

EXPLORATION:

- The teacher will define what each instrument measures:
  Anemometer: any instrument for measuring the speed of wind.
  Barometer: any instrument that measures atmospheric pressure.
  Rain Gauge: an instrument for measuring rainfall.

EXPLANATION:

- Teacher will show the students how to read the instruments and how the information is used to analyze and interpret weather conditions just as the thermometer is used to tell us the temperature.
- Students will record the information from each instrument daily or weekly to collect measureable data.
- Student will maintain a weather journal in which they record specific data from each instrument. The journal should include the date, time, and accurate measurement.
- Vocabulary: anemometer, barometer, rain gauge, wind gust, pressure, dew points.

ELABORATION:

- Students can discuss which instrument is most helpful at school and for the garden.
- Students will maintain the weather instruments that remain in the garden area throughout the year.
- On a regular basis students will complete a *Weather Instruments* student journal page.

EVALUATION:

- Students will compile the *Weather Instruments* student journal pages in a binder.
- Teacher will observe reactions and interactions with other students and the weather instruments.

WEATHER INSTRUMENTS JOURNAL PAGE

Name ______________________________

Date ____________________ *Time* ____________________

Temperature

____

Barometric Pressure

____

Wind Speed

____

Describe the weather in the garden today.

____

____

____

____

____

SECOND GRADE LESSON PLAN
Who Likes to Eat Worms?

SUBJECT / GRADE LEVEL: Second Grade - Life Science - Animals and their Environment.

ACADEMIC STANDARDS AND PERFORMANCE INDICATORS:

ELA:

W.MCC.2.2 Plan, revise, and edit, focusing on a topic while building on personal ideas and the ideas of others to strengthen writing.

W.L.4.8 Produce, expand, and rearrange complete simple and compound sentences.

C.MC.1.4 Participate in shared conversations with varied partners about focused grade level topics and texts in small and large groups.

RL.MC.5.1 Ask and answer literal and inferential questions to demonstrate understanding of a text; use specific details to make inferences and draw conclusions in texts heard or read.

RL.LCS.10.1 Use context to determine the meaning of words and phrases.

RI.LCS.9.4 Use print and multimedia resources to determine or clarify the precise meaning of words and phrases.

Science:

2.L.5A.1 Obtain and communicate information to classify animals such as mammals, birds, amphibians, reptiles, fish, or insects based on their physical characteristics.

2.L.5A.2 Construct explanations for how structures including structures for seeing, hearing, grasping, protection, locomotion, and obtaining and using resources of different animals help them survive.

2.L.5A.3 Construct explanations using observations and measurements of an animal as it grows and changes to describe the stages of development of the animal.

Math:

2.MDA.1 Select and use appropriate tools e.g., rulers, yardsticks, meter sticks, measuring tapes to measure the length of an object.

2.MDA.2 Measure the same object or distance using a standard unit of one length and then a standard unit of a different length and explain verbally and in writing how and why the measurements differ.

2.MDA.3 Estimate and measure length/distance in customary units i.e., inch, foot, yard and metric units i.e., centimeter, meter.

2.MDA.4 Measure to determine how much longer one object is than another, using standard length units.

2.MDA.10 Draw conclusions from T-Charts, object graphs, picture graphs, and bar graphs.

LESSON OBJECTIVES:

TSW identify the needs of the earthworms as well as their place in the food web.

MATERIALS:

- Earthworms
- Containers for students to see the earthworm, plastic boxes or cups
- *Wiggling Worms at Work* by Wendy Pfeffer
- *Who Likes To Eat Worms?* student investigation guide (attached)
- *Who Likes To Eat Worms?* teacher guide (attached)

ENGAGEMENT:

- Teacher will pose the question, "**What does a worm eat for breakfast?**"
- Teacher will follow the engagement question by reading *Wiggling Worms at Work* by Wendy Pfeffer to the class.

EXPLORATION:

- Students will observe earthworms in a container and discuss what animals may eat these worms in nature.

EXPLANATION:

- Teacher will explain the importance of worms in the food web.

ELABORATION:

- Students write a poem about worms as they complete their *Who Likes To Eat Worms?* student investigation guide.

EVALUATION:

- Students will turn in their *Who Likes To Eat Worms?* student investigation guide.
- Teacher will observe reactions and interactions with other students and the worms.

RESOURCES:

1. *Wiggling Worms at Work* by Wendy Pfeffer.
2. *Who Likes to Eat Worms?* handout (attached).

TEACHER GUIDE

WHO LIKES TO EAT WORMS?

BACKGROUND:

Students should understand that worms are an important part of the food web.

ENGAGEMENT:

- Teacher will pose the question, "**What does a worm eat for breakfast?**"
- Teacher will follow the engagement question by reading *Wiggling Worms at Work* by Wendy Pfeffer to the class.

EXPLORATION:

- Students will observe the earthworms in the containers.
- Students come up with a list of what they think a worm needs to live.
- Students look at pictures of animals and discuss which animals might eat earthworms.

EXPLANATION:

- Students discuss what they think the worm eats and needs to survive.
- Earthworms are important members of the food web.
- They serve as food for many kinds of animals. For example, moles tunnel through the soil in search of earthworms. The moles catch the worms and place them in a storage area for later consumption. To keep a worm from escaping, the mole may bite off its first three to five anterior segments.
- The worm has needs as well. They need damp soil, a mild climate, and the availability of decaying organic material.
- **If there were no worms would there be a difference in the environment? What differences may there be?**
- **If a worm loses one body part will it regenerate? What does this mean for the worm s survival?**
- **Can worms survive without soil?**
- **How might an earthworm be important to a garden ecosystem?**
- Vocabulary words: *decaying*, *organic*, *cinquain*, *anterior*, *regenerate*.

ELABORATION:

- Students complete food chains, the link of life of the earthworm—students can make food chain starting with a part of a plant.
- The teacher writes a cinquain poem about worms with the students.
  ***A cinquain is a five line poem that does not rhyme.
- Students write a poem about worms as they complete their *Who Likes To Eat Worms?* student investigation guide.

EVALUATION:

- Students will turn in their *Who Likes To Eat Worms?* student investigation guide.
- Teacher will observe reactions and interactions with other students and the worms.

WHO LIKES TO EAT WORMS?

Name

Draw a picture of the food web including worms.

Write a cinquain poem about worms.

LESSONS FOR THIRD GRADE

THIRD GRADE LESSON PLAN
Disappearing Banana Peel

SUBJECT / GRADE LEVEL: Third Grade - Inquiry and Life Science.

ACADEMIC STANDARDS AND PERFORMANCE INDICATORS:

Science:

3.L.5A.2 Develop and use a food chain model to classify organisms as producers, consumers, and decomposers and to describe how organisms obtain energy.

3.S.1A.1 Ask questions that can be 1.) answered using scientific investigations or 2.) used to refine models, explanations, or designs.

3.S.1A.3 Plan and conduct scientific investigations to answer questions, test predictions and develop explanations: 1.) formulate scientific questions and predict possible outcomes, 2.) identify materials, procedures, and variables, 3.) select and use appropriate tools or instruments to collect qualitative and quantitative data, and 4.) record and represent data in an appropriate form. Use appropriate safety procedures.

3.S.1A.4 Analyze and interpret data from observations, measurements, or investigations to understand patterns and meanings.

3.S.1A.6 Construct explanations of phenomena using 1.) scientific evidence and models, 2.) conclusions from scientific investigations, 3.) predictions based on observations and measurements, or 4.) data communicated in graphs, tables, or diagrams.

3.S.1A.7 Construct scientific arguments to support claims, explanations, or designs using evidence from observations, data, or informational texts.

LESSON OBJECTIVES:

TSW explain the process of decomposition and justify its importance in a food chain. TSW observe the decomposition process and analyze what is occurring. TSW construct reasonable explanations for how the decomposition process occurs. ****This lesson works well with the Renewable and Nonrenewable Resources Lesson pg. 185.*

MATERIALS:

- Paper plate
- Banana and plastic bottle for each group or pair
- *Disappearing Banana Peel* student investigation guide (attached)
- *Disappearing Banana Peel* teacher guide (attached)

ENGAGEMENT: IN THE CLASSROOM OR GARDEN

- Teacher will pose the question, "**How would a banana peel disappear over time?**"
- Teacher will show the following video to the class: *How compost is made* by Recycle Now Campaign: https://www.youtube.com/watch?v=cBkBwVFFEWw.

EXPLORATION: IN THE CLASSROOM

- Teacher will introduce the concept of decomposition.

EXPLANATION:

- Teacher will facilitate the setup of the banana peel and water bottle decomposition experiment.

ELABORATION: IN THE CLASSROOM OR GARDEN

- Teacher will take students out to the compost bin.
- Teacher will use inquiry based questions to conduct a class discussion in order for students to investigate the process of decomposition.
- Students will practice their observation and illustration skills while completing their *Disappearing Banana Peel* student investigation guide.

EVALUATION:

- Students will turn in the *Disappearing Banana Peel* student investigation guide.
- Teacher will observe reactions and interactions with other students and the banana peel.

RESOURCES:

1. *How compost is made* by Recycle Now Campaign: https://www.youtube.com/watch?v=cBkBwVFFEWw.
2. *Disappearing Bannana Peel* handout (attached).

TEACHER GUIDE
DISAPPEARING BANANA PEEL

BACKGROUND:

Student should understand that organic items will break down over time due to the process of decomposition.

ENGAGEMENT:

- Teacher will pose the question, "**How would a banana peel disappear over time?**"
- Teacher will show the following video to the class: *How compost is made* by Recycle Now Campaign: https://www.youtube.com/watch?v=cBkBwVFFEWw.

EXPLORATION:

****This lesson could be used before or after an introduction lesson on food chains producers, decomposers and consumers.*

****This activity will take place over several days as students observe the decomposition of the banana peel.*

- Take students to the garden to observe the compost bin and/or worm bin. Pose questions regarding the compost bin: **What do you think is happening in the bin? How is it happening? What is the compost being used for? What does it provide?**
- The teacher will introduce today's focus question: **What is decomposition? How does it occur?**

EXPLANATION:

- Decomposition is the breakdown or recycling of organic matter into smaller components that can be used by other organisms.
- Decomposers are important organisms that decompose the remaining waste of other organisms, dead organic matter, and some inorganic matter. The smallest—but most important—decomposers are the microscopic bacteria and fungi. Other larger decomposers include lichen, mosses, worms, and wood eating insects such as termites, beetles, and certain ants. Insects and their larvae bore through decomposing matter creating tunnels and spaces, which aerates the compost, allowing the bacteria and fungi deeper access within the material.

- Decomposing matter also can provide homes and food for many organisms. Many factors contribute to the processes of decomposition, including temperature, amount of light, aeration, moisture, type of material that is decomposing, and the source of bacteria and fungi.
- Generally, higher moisture and higher temperature can accelerate the rate of decomposition.
- **What are your predictions for what will happen to the banana? The water bottle? Why do you predict that will happen? What evidence or prior knowledge do you have to support your prediction?**

  BREAK UNTIL THE FOLLOWING DAY: Observe for at least 3 days. The banana peel will change quickly.
- On the third day, ask students to describe what they think is happening to the banana peel. **Why did the appearance of the banana peel change but not the appearance of the water bottle? Describe the differences in the material of the banana peel and the water bottle?**
- Ask students to name things that they think are organic, and things that they think are inorganic. After taking suggestions, tell students that things that are organic can decompose, while some things that are inorganic take a longer time to decompose or may not decompose at all. **Is the banana peel organic or inorganic? What about the bottle?**
- Ask the students to think about how the decomposition of the banana peel relates to worms.
- After taking suggestions, tell students that invisible microorganisms, such as bacteria, help break down the banana peel, similar to the way that worms help break down organic matter.
- Students will begin *Disappearing Banana Peel* student investigation guide.

ELABORATION:

- Ask students, **What should we do with the banana peel now? What have we learned about how living things obtain energy? Where do plants get their energy from?**
- If not suggested by a student, the teacher should suggest adding the banana peels to the garden compost bins. Lead a classroom discussion about how this banana peel will provide energy for decomposers which will make the soil rich for our plants.
- Also lead a discussion about the plastic bottle. **So, what happens to the bottle? How could this pose a problem to our planet? What could or should we do as humans to solve this problem?**
- Other discussion topics or possible extensions:

  –**Why are decomposers important?**

  –**What would happen if we didn t have decomposers?**

  –**Would covering the banana peel with plastic wrap affect its rate of decomposition? Why/why not?**

  –**How can we set up an experiment to determine how quickly different organic materials take to decompose?**

- Students will complete *Disappearing Banana Peel* student investigation guide.

EVALUATION:

- Students will turn in the *Disappearing Banana Peel* student investigation guide.
- Teacher will observe reactions and interactions with other students and the banana peels.

DISAPPEARING BANANA PEEL

Name

___

Draw a picture of each object.

Banana	*Plastic Bottle*

Predict what will happen when we leave these items for several days.

Banana	*Plastic Bottle*

Describe the items at the end of one week.

Banana	*Plastic Bottle*

THIRD GRADE LESSON PLAN
Renewable and Nonrenewable Resources

SUBJECT / GRADE LEVEL: Third Grade - Earth Science - Renewable and Nonrenewable Resources.

ACADEMIC STANDARDS AND PERFORMANCE INDICATORS:

Science:

3.E.4A.3 Obtain and communicate information to exemplify how humans obtain, use, and protect renewable and nonrenewable Earth resources.

3.E.4B.4 Define problems caused by a natural event or human activity and design devices or solutions to reduce the impact on the environment.

LESSON OBJECTIVES:

TSW explain the difference between renewable and nonrenewable resources; TSW identify examples of each. TSW justify the importance of reducing, reusing and recycling to help maintain Earth's supply of nonrenewable resources.

MATERIALS:

- *The Lorax* by Dr. Seuss
- *Renewable and Nonrenewable Resources* student investigation guide (attached)
- *Renewable and Nonrenewable Resources* teacher guide (attached)

ENGAGEMENT:

- Teacher will pose the question, "**Are plants a renewable resource?**"
- Teacher will follow the engagement question by reading *The Lorax* by Dr. Seuss to the class.

EXPLORATION:

- Teacher will introduce the terms *renewable* and *nonrenewable*.
- The teacher will make a T-Chart on the board without labels: On one side begin listing natural/ renewable resources (plants, animals, air, water, sunlight, etc.) and on the other side begin listing nonrenewable resources (plastic, aluminum, coal and fossil fuels).

EXPLANATION:

- Teacher will take students to the garden to begin their *Renewable and Nonrenewable Resources* student investigation guide.

ELABORATION:

- The class can brainstorm problems with nonrenewable resources in the school including use of plastic water bottles, newspapers piling up, etc...
- The class can think of solutions to these problems.
- The class can create a campaign to present to the school for ways to solve and why this is important.

EVALUATION:

- Students will turn in the *Renewable and Nonrenewable Resources* student investigation guide.
- Teacher will observe reactions and interactions with other students and the renewable and nonrenewable resources.

RESOURCES:

1. *The Lorax* by Dr. Seuss.
2. *Renewable and Non-Renewable Resources* handout (attached).

TEACHER GUIDE

RENEWABLE AND NONRENEWABLE RESOURCES

BACKGROUND:

Students should understand that there is a difference in renewable and nonrenewable resources.

ENGAGEMENT:

- Teacher will pose the question, "**Are plants a renewable resource?**"
- Teacher will follow the engagement question by reading *The Lorax* by Dr. Seuss to the class.

EXPLORATION:

- The teacher will make a T-Chart on the board without labels. On one side begin listing natural/ renewable resources (plants, animals, air, water, sunlight, etc.) and on the other side begin listing nonrenewable resources (plastic, aluminum, coal and fossil fuels).
- Students think-pair-share what they think the labels should be for each.
- Challenge students to put the Truffula Trees on the correct side of the T-Chart.
- Introduce terms "renewable" and "nonrenewable."

EXPLANATION:

- Teacher will take students to the garden. Students will begin their *Renewable and Nonrenewable Resources* student investigation guide.
- Students should list things that they see, placing them under the correct category. Students will be given time to adjust their charts after returning to the classroom for direct instruction. ****Optional Alternative: If applicable, students can use tablets / iPads to take pictures of renewable/ nonrenewable items in the garden. Students can use pictures to create presentations to share with the class.*
- Explain the definitions of renewable and nonrenewable.
- Renewable: Resources that can be replaced naturally or through human assisted actions.
- Nonrenewable: Natural resources that are available in limited amounts or take millions of years to replace.

ELABORATION:

- Literature tie-in: **Were the Truffula Trees renewable or nonrenewable? What could the Oncler have done to save the Truffula Trees?**

- Teacher can pass out magazines to students and have them cut, sort and create a poster of renewable and nonrenewable.
- The class can brainstorm problems with nonrenewable resources in the school: use of plastic water bottles, newspapers piling up, etc. and think of solutions to these problems. The class can create a campaign to present to the school for ways to solve and why this is important.
- Newspapers can be used to line garden beds to prevent weeds growing.
- Plastic bottles can be used to grow seedlings by cutting off the tops.
- Students will complete *Renewable and Nonrenewable Resources* student investigation guide.

EVALUATION:

- Students will turn in the *Renewable and Nonrenewable Resources* student investigation guide.
- Teacher will observe reactions and interactions with other students and the renewable and nonrenewable resources.

RENEWABLE AND NONRENEWABLE RESOURCES

Name

____

Make a list of renewable and nonrenewable resources.

Renewable Resources	*Nonrenewable Resources*

Define a renewable resource.

____

____

Define a nonrenewable resource.

____

____

THIRD GRADE LESSON PLAN
Engineering in the Garden

SUBJECT / GRADE LEVEL: Third Grade - Engineering in the Garden - Beneficial and Harmful Insects and Organisms.

ACADEMIC STANDARDS AND PERFORMANCE INDICATORS:

Science:

3.L.5B.1 Obtain and communicate information to explain how changes in habitats such as those that occur naturally or those caused by organisms can be beneficial or harmful to the organisms that live there.

3.S.1B.1 Construct devices or design solutions to solve specific problems or needs: 1.) ask questions to identify problems or needs, 2.) ask questions about the criteria and constraints of the devices or solutions, 3.) generate and communicate ideas for possible devices or solutions, 4.) build and test devices or solutions, 5.) determine if the devices or solutions solved the problem and refine the design if needed, and 6.) communicate the results.

3.E.4B.4 Define problems caused by a natural event or human activity and design devices or solutions to reduce the impact on the environment.

LESSON OBJECTIVES:

TSW research the insects that are beneficial and harmful to the garden. Students will design, test, and improve modify solutions to attract beneficial insects and to deter harmful insects from the garden.

MATERIALS:

- *Engineering in the Garden* student investigation guide (attached)
- *Engineering in the Garden* teacher guide (attached)

ENGAGEMENT: IN THE CLASSROOM

- Teacher will pose the question, "**What is a pest?**"

EXPLORATION:

- Teacher will introduce the design challenge in which each team must design, test, and improve a strategy to attract beneficial insects to the garden and deter harmful insects from the garden.

EXPLANATION:

- Student teams will determine what strategies a process or a system they want to implement into the garden.
- Students may illustrate a model, list materials and/or save images of the strategies/models they will create.
- Teacher will rotate from team to team asking questions about the different strategies that each team has chosen.
- Students will begin *Engineering in the Garden* student investigation guide.

ELABORATION:

- Students will share their pest management strategies will the class before implementing them in the garden.
- Students will complete their *Engineering in the Garden* student investigation guide.

EVALUATION:

- Students will turn in their *Engineering in the Garden* student investigation guide.
- Teacher will observe reactions and interactions with other students and the insects.

RESOURCES:

1. *Engineering in the Garden* handout (attached).

TEACHER GUIDE
ENGINEERING IN THE GARDEN

BACKGROUND:

Students should understand that some insects are harmful to garden plants while others are beneficial.

ENGAGEMENT:

- Teacher will pose the question, "**What is a pest?**"

EXPLORATION:

- The wonderful insects that dine only on landscape pests and don't harm good bugs or plants are called *beneficial insects* because they are helpful to our garden. These are the bugs gardeners and farmers want to attract.
- There are other bugs which gardeners and farmers consider to be a nuisance. These pests are harmful because they feed on plants and can ruin entire crops. These are the bugs gardeners and farmers want to repel or deter from the garden.

PRESENT THE CHALLENGE:

- Each team must design, test, and improve a strategy to attract beneficial insects to the garden and deter harmful insects from the garden.

CRITERIA:

***Invite students to ask as many questions as they can about the criteria. Student inquiries often lead to important revisions to the list that are unforeseen by the teacher.

- Must attract at least one beneficial insect to the garden that does not currently use the garden as a habitat.
- Must deter at least one harmful insect that currently resides in the garden.
- Attractants and deterrents may be a scent, a process, or something physical.
- Must use materials that are readily available at school or home.
- May not purchase a pesticide solution.
- All strategies must use repurposed, recycled materials approved a facilitator.

EXPLANATION:

- Student teams will research, discuss, observe, and record the existing beneficial and harmful insects that reside in the garden. Next, they will determine which bug they would like to attract and which bug they would like to deter based on the current needs of the garden.

- Student teams will determine what strategies a process or a system they want to implement into the garden. Students may illustrate a model, list materials, and/or save images of the strategies/models they will create.
- Student teams will assemble their strategies. While students plan and create, the teacher should be asking student teams questions to facilitate the task. Raise questions to helps students consider things they did not notice, such as, "**Can you explain why your team decided this piece is necessary?**"

DETERRENTS: SLUGS

- Use a fruit trap. Slugs are naturally attracted to fruits and the plants in your garden, so use that attraction to your advantage to trap them. Place a board out near where the slugs have been eating your plants, and put half of an orange or a few slices of cabbage "slug favorites" in the center. Leave the board out overnight, and in the morning the slugs should be covering the morsels of food you left on the board. Relocate any slugs to a safe habitat.

ATTRACTANTS: LADYBUGS

- Mash up non-acidic fruit such as raisins, prunes, or blueberries and add to various shaped homes tubes, boxes, etc...
- Students may also engineer preventative solutions such as physical barriers to deter all kinds of organisms.
- Students will begin their *Engineering in the Garden* student investigation guide.

ELABORATION:

- Student teams will share their strategies/models with the classroom community before incorporating them into the garden.
- Students will explain their reasoning and answer questions that arise from other student teams.
- Students may make adjustments based on the feedback given from their peers.
- Students may analyze qualitative or quantitative observation data to determine the success of their strategies/models. They may keep a log with the time, date, location and an estimated number of the insects they see in the garden.
- Student teams will test and improve, then test and improve, and continue this cycle based on their findings over several weeks. Students need to analyze the variables that may be causing their strategy to be unsuccessful.
- Students will complete their *Engineering in the Garden* student investigation guide.

EVALUATION:

- Students will turn in their *Engineering in the Garden* student investigation guide.
- Teacher will observe reactions and interactions with other students and the insects.

ENGINEERING IN THE GARDEN

Name ____

Describe the beneficial insect that you will be attracting to the garden as well as the strategy that you will use to attract it.

____

____

____

____

Draw a picture of your solution.

Describe the pest that you will be keeping out of the garden as well as the strategy that you will use to keep it away.

____

____

____

____

Draw a picture of your solution.

THIRD GRADE 5E LESSON PLAN

Peek Inside the Soil

SUBJECT / GRADE LEVEL: Third Grade - Earth Science.

ACADEMIC STANDARDS AND PERFORMANCE INDICATORS:

Science:

3.E.4A.1 Analyze and interpret data from observations and measurements to describe and compare different Earth materials including rocks, minerals, and soil and classify each type of material based on its distinct physical properties.

LESSON OBJECTIVE: TSW closely observe soil samples from around the school to determine the components of soil.

MATERIALS:

- Three trays or flat containers, aluminum baking sheets
- Trowel
- Hand lenses
- Clear tube or jar (optional for extension investigation)
- *A Handful of Dirt* by Raymond Bial
- *Peek Inside the Soil* student investigation guide (attached)
- *Peek Inside the Soil* teacher guide (attached)

ENGAGEMENT:

- Teacher will pose the question, "**If you were going to cook up a batch of soil, what ingredients would you use?**"
- Teacher will follow the engagement question by reading A *Handful of Dirt* by Raymond Bial to the class.

EXPLORATION:

- Teacher will guide students in creating a chart of qualitative observations based on the components of soil.

EXPLANATION:

- Teacher will take students around the campus to collect three different soil samples.
- Students will begin their *Peek Inside the Soil* student investigation guide.

ELABORATION:

- Teacher will encourage students to explore the components of soil separated out individually.
- Students will complete the *Peek Inside the Soil* student investigation guide.

EVALUATION: IN THE CLASSROOM

- Students will turn in the *Peek Inside the Soil* student investigation guide.
- Teacher will observe reactions and interactions with other students and the soil samples.

RESOURCES:

1. *A Handful of Dirt* by Raymond Bial.
2. *Peek Inside the Soil* handout (attached).
3. *Soil Observations* Student Example (attached).

TEACHER GUIDE
PEEK INSIDE THE SOIL

BACKGROUND

Students should understand that soil has many components including water, air, organic matter, sand, silt and clay.

ENGAGEMENT:

- Teacher will pose the question, "**If you were going to cook up a batch of soil, what ingredients would you use?**"
- Teacher will follow the engagement question by reading *A Handful of Dirt* by Raymond Bialto the class.

EXPLORATION:

- **What do you think makes up soil?** Reintroduce how to make qualitative observations. It would be helpful to display an anchor chart that has words to describe properties. For example:
  Texture: moist, dry, smooth, rough, coarse, fine.
  Shape: circular, flat, holey, rigid.
  Color: light, dark, brown, green.
  Smell: clean, earthy, rotten, sweet.

EXPLANATION:

- Explain collection of soil samples procedure.
- Students will go to three different locations to collect soil samples. Make sure to dig deep enough to get lots of soil types in the samples.
- Students rotate to three different tables to make close observations of soil samples and compare the samples using hand lenses.
- Allow five minutes to share observations. **Is all soil the same? Why does it look and feel different in different places? What makes up soil?**
- Students will begin their *Peek Inside the Soil* student investigation guide.
- Vocabulary: *particle*, *physical properties*, *texture*, *sample*, *soil composition*.

ELABORATION:

- **What would we find if we separated all of the things that make up soil?** Add soil from all three locations to a clear tube or jar. While students watch, pour water into the tube and ask them to consider why there are air bubbles. Explain that soil is also made up of air.
- Shake the samples vigorously and ask students to predict what will happen if we let the tube or jar sit overnight undisturbed.
- Students will complete *Peek Inside the Soil* student investigation guide.

EVALUATION:

- Students will turn in the *Peek Inside the Soil* student investigation guide.
- Teacher will observe reactions and interactions with other students and the soil samples.

PEEK INSIDE THE SOIL

Name

____

Describe the three soil samples that you collected.

Soil Sample #1	
Soil Sample #2	
Soil Sample #3	

What would we find if we separated all of the things that make up soil?

____

____

____

____

9

Soil observations

8:57am
12/9/2013

Sample 1 / Swing Area

black minarals feels like sand It is brownish finey rocks It has wood chips

Sample 2 / by the garden

It has rocks and grass, roots white balls, clay.

Sample 3 / by the sidewalk

It got hard it has clay there are wood chips chuncks of clay and soil.

THIRD GRADE LESSON PLAN
Pollinator Habitats

SUBJECT / GRADE LEVEL: Third Grade - Life Science - Pollinator Habitats.

ACADEMIC STANDARDS AND PERFORMANCE INDICATORS:

Science:

3.S.1 The student will use the science and engineering practices, including the processes and skills of scientific inquiry, to develop understandings of science content.

4.S.1 The student will use the science and engineering practices, including the processes and skills of scientific inquiry, to develop understandings of science content.

3.L.5A.1 Analyze and interpret data about the characteristics of environments including salt and fresh water, deserts, grasslands, forests, rain forests, and polar lands to describe how the environment supports a variety of organisms.

4.L.5B.2 Construct explanations for how structural adaptations such as the types of roots, stems, or leaves; color of flowers; or seed dispersal allow plants to survive and reproduce.

LESSON OBJECTIVES:

TSW discuss the definition, function and importance of pollination. TSW discuss pollinator habitats and the ways in which humans affect pollinator/plant relationships in negative and positive ways. TSW analyze data to predict how many pollinators will visit the garden.

MATERIALS:

- Three index cards pre-labeled by teacher with insect name and the flower colors that attract the insect
- Large pictures of a variety of colored flowers for questioning
- *Pollinator Habitats* student investigation guide (attached)
- *Pollinator Habitats* teacher guide (attached)

ENGAGEMENT:

- Teacher will pose the question, "**Which pollinators are likely to visit and pollinate our garden most often?**"

- Teacher will follow the engagement question by showing this video to the class: *The Beauty of Pollination - Moving Art* by Louie Schwartzberg: https://www.youtube.com/watch?v=MQiszdkOwuU.

EXPLORATION:

- Teacher will divide students into groups and take them to the garden to search for pollinators.

EXPLANATION:

- Using the *Pollinator Habitats* teacher guide, the teacher will guide students through the completion of the *Pollinator Habitats* student investigation guide.

ELABORATION:

- Students will complete *Pollinator Habitats* student investigation guide.
- Students can predict which pollinators will be attracted to the garden most often based on the flower colors that are more prevalent in the garden; to make a good prediction, students will need to find the sum of the flower colors that attract each pollinator.

EVALUATION:

- Students will turn in the *Pollinator Habitats* student investigation guide.
- Teacher will observe reactions and interactions with other students and the pollinators.

RESOURCES:

1. *The Beauty of Pollination - Moving Art* by Louie Schwartzberg: https://www.youtube.com/watch?v=MQiszdkOwuU.
2. *Polinator Habitats* handout (attached).

TEACHER GUIDE
POLLINATOR HABITATS

BACKGROUND:

Students should understand that there are many types of pollinators that play an important role in the garden.

ENGAGEMENT:

- Teacher will pose the question, "**Which pollinators are likely to visit and pollinate our garden most often?**"
- Teacher will follow the engagement question by showing this video to the class: *The Beauty of Pollination - Moving Art* by Louie Schwartzberg: https://www.youtube.com/watch?v=MQiszdkOwuU.

EXPLORATION:

- Students will be divided into three groups. Each group will be a different kind of pollinator: Bees, Butterflies or Flies there are other pollinators, but we'll focus on these three for now.
- Each group will receive an index card that describes the colors of flowers that attract their pollinator. The index cards may be labeled as follows with space for students to make tally marks.
  Bees: yellow/blue/purple
  Butterflies: red/orange/yellow/pink/blue
  Flies: green/white/cream
- Instruct students to walk through the entire garden and tally the number of places where they find the colors or types of flowers their pollinator would be likely to visit.
- Return to an outdoor or indoor classroom when students have counted and recorded the data. Record the colors that attract each pollinator and student findings on the board.
- Hold up or display large pictures of flowers and question students about which pollinator would be attracted to these flowers.
- Students will use the data to make a bar graph that displays the number of plants in the garden that represent different colors.

EXPLANATION:

Guide students through the completion of their *Pollinator Habitats* student investigation guide using the information below:

- **What is pollination?** Pollination is the fertilization process of a plant which leads to production of fruit and seed. During the process, pollen is moved from one flower to another plant or to another part of the same plant. When pollen from the stamen comes in contact with the pistil of the flower, the flower produces seeds.
- **Why is pollination important?** Pollination is what ensures that plants do not go extinct, especially those which provide us with food. Pollinators are what make our meals possible. One of every three bites of food eaten worldwide depends on pollinators. Honeybees are responsible for 75 percent of that pollination.
- **Besides honeybees, who are the pollinators?** Pollinators, which include thousands of insect species, include native bees most bees are not social, they live in woodpecker holes and other small burrows they find in the wild, tiny wasps, butterflies, beetles, flies, and other animals, such as hummingbirds and bats.
- FUN FACT: Chocolate plants are "only" pollinated by a small fly called a midge.

HOW DO HUMANS AFFECT POLLINATORS?

- Plant scientists are concerned about our role in weakening pollinator/plant relationships.
- The overuse of pesticides, which often kill beneficial pollinators, is one factor. Another one, particularly serious for migrating pollinators such as monarch butterflies, is cities/land fragmentation that results largely from development. Isolated plants can t attract a variety of pollinators or visitors frequent enough to sustain the plants and ultimately their partners. There was a time when native bees and wild honey bees performed all of a farmer's pollination needs because of the presence of natural areas nearby. These natural areas provided nesting sites, food and protection for the bees but are being developed rapidly.
- Questioning: "**What kinds of things do humans develop which may replace these natural habitats?**" If we understand pollinators and how plants attract them, we can create a habitat that will attract them to our garden.
- Questioning: "**What does every good habitat provide organisms?**" Students: "air, water, sunlight, nutrients, and space."
- Questioning: "**What are the primary foods that attract pollinators to flowers?**" Students: "nectar and pollen". Nectar has carbohydrates for energy. Pollen is full of protein and nutrients.
- Vocabulary: *attract*, *pollen*, *nectar*, *pollinator*, *habitat*, *reproduce*.

ELABORATION:

- Students can predict which pollinators will be attracted to the garden most often based on the flower colors that are more prevalent in the garden. To make a good prediction, students will need to find the sum of the flower colors that attract each pollinator.
- Students may benefit from using a teacher provided sentence starter such as "I predict that the pollinators that are likely to visit our garden the most often are. . . I predict this because. . . ."
- Students will complete *Pollinator Habitats* student investigation guide.

EVALUATION:

- Students will turn in the *Pollinator Habitats* student investigation guide.
- Teacher will observe reactions and interactions with other students and the pollinators.

POLLINATOR HABITATS

Name

___

What is pollination?

___

___

How do humans affect pollinators?

___

___

___

List five different pollinators.

THIRD GRADE LESSON PLAN
Worms at Work

SUBJECT / GRADE LEVEL: Third Grade - Life Science.

ACADEMIC STANDARDS AND PERFORMANCE INDICATORS:

Science:

3.L.5B.2 7 Develop and use models to explain how changes in a habitat cause plants and animals to respond in different ways such as hibernating, migrating, responding to light, death, or extinction.

LESSON OBJECTIVES:

TSW investigate the body parts of worms as well as their reactions to stimuli such as light, vinegar, shelter rocks in their habitat soil.

MATERIALS:

- Worms (earthworms can be collected on your school site or red wigglers can be purchased online or at a local bait store)
- Stones
- Pencil for each student
- Clear plastic plates or petri dishes
- Flashlights
- Spray bottle to keep worms moist during investigation
- Strawberries
- Pickles
- Hand lenses
- *Worms at Work* student investigation guide (attached)
- *Worms at Work* teacher guide (attached)

ENGAGEMENT:

- Teacher will pose the question, "**What do you know about worms?**"

EXPLORATION:

- Teacher will divide students into groups and give each group a worm and plate for the experiment.

EXPLANATION:

- Each group will expose their worms to the strawberry, pickle, stone and flashlight.
- Each group will record the reaction of the worms after each exposure on their *Worms at Work* student investigation guide.

ELABORATION:

- After the experiment is over, the class can build a worm bin to house the worms and make vermicompost.
- If the class does not build a worm bin, worms can be released into the soil in the garden.
- Students will complete *Worms at Work* student investigation guide.

EVALUATION: IN THE CLASSROOM

- Students will turn in the *Worms at Work* student investigation guide.
- Teacher will observe reactions and interactions with other students and the worms.

RESOURCES:

1. *Worms at Work* handout (attached).

TEACHER GUIDE
WORMS AT WORK

BACKGROUND:

Students should understand that worms play an important role in the garden. Students should understand that worms react to different types of stimuli in the garden.

ENGAGEMENT:

- Teacher will pose the question, "**What do you know about worms?**"

EXPLORATION:

- **How do worms use their bodies and react to changes in their habitat?**
- Provide each group of learners at least two worms and a clump of moist soil on a clear plastic plate or petri dish. Spray the worms as needed so they don't lose body moisture during the activity.
- Students will need a strawberry, a pickle a flashlight and a stone.

EXPLANATION:

- Each group will expose their worm to the strawberry, pickle, stone and flashlight. After each item, students should take time to record the reaction of the worm on their *Worms at Work* student investigation guide.

ELABORATION:

- After the experiment is over, the class can build a worm bin to house the worms and make vermicompost.
- If the class does not build a worm bin, worms can be released into the soil in the garden.
- Students will complete *Worms at Work* student investigation guide.

EVALUATION:

- Students will turn in the *Worms at Work* student investigation guide.
- Teacher will observe reactions and interactions with other students and the worms.

WORMS AT WORK

Name

___

Why do we need to mist the worms with water?

___

___

___

How did your worm react to each item below?

Light	
Strawberry	
Pickle	
Stone	

THIRD GRADE LESSON PLAN

Go Away Weeds

SUBJECT / GRADE LEVEL: Third Grade - Life Science.

ACADEMIC STANDARDS AND PERFORMANCE INDICATORS:

Science:

3.L.5B.2 Obtain and communicate information to explain how changes in habitats such as those that occur naturally or those caused by organisms can be beneficial or harmful to the organisms that live there.

LESSON OBJECTIVES:

TSW will explore the resources that weeds can take from garden plants such as sunlight, water and nutrients. TSW identify and remove unwanted weeds from the garden.

MATERIALS:

- Buckets for weeds (one for each group)
- *Go Away Weeds* student investigation guide (attached)
- *Go Away Weeds* teacher guide (attached)

ENGAGEMENT:

- Teacher will pose the question, "**What is a weed?**"

EXPLORATION:

- Teacher will conduct an exercise in which students will pretend to be weeds in the garden competing for the resources that are used by garden plants as well.

EXPLANATION:

- Teacher will provide student teams with buckets for weed collection as well as instructions for proper weed removal.
- Student teams will compete to fill their bucket first.

ELABORATION:

- Students will determine the best disposal method for their weeds.
- Students will complete their *Go Away Weeds* student investigation guide.

EVALUATION:

- Students will turn in their *Go Away Weeds* student investigation guide.
- Teacher will observe reactions and interactions with other students and the weeds.

RESOURCES:

1. *Go Away Weeds* handout (attached).

TEACHER GUIDE
GO AWAY WEEDS

BACKGROUND:

Students should understand that weeds can compete with garden plants for important resources.

ENGAGEMENT:

- Teacher will pose the question, "**What is a weed?**"

EXPLORATION:

- A *weed* is classified as a plant that is growing where it is not wanted. Weeds compete with the desired plant for resources such as light, nutrients, and space. They are known to interfere with food production. Also known as *invasive plants*, weeds are fast-growing and have few predators. Pulling weeds is much easier when the ground is wet, so pulling after a heavy rain will make the job go more smoothly.
- Invite students to stand shoulder to shoulder and ask them if they could eat, sleep, and move with only this amount of space. Explain that like humans, plants need space to move and grow. Explain that some of the plants in the garden are currently fighting for space because they are being invaded by weeds. When the weeds move into the space of our desired plants, they compete for sunlight, nutrients, and water.

EXPLANATION:

- After providing each team with a bucket to collect the wanted weeds, the teacher should demonstrate how to properly identify and pull some of the weeds without disrupting the root system of the desired plants.

STEPS:

- Grab the weed at the base of the largest stem as far down as possible.
- Grip the weed tightly, wiggle it a bit, and pull sharply.
- If working with stubborn weeds, you may need to use a pointed tool to loosen the soil around the weeds before pulling.
- Next, give students a time limit, and allow teams to get started it may be helpful to have some other adults to monitor and guide the pulling to ensure students aren't pulling the wrong plants.

CHALLENGE:

- Students will be put into groups and challenged to fill their bucket first with the most weeds from the garden.

ELABORATION:

- If most of weeds are in the adult stage of their life cycle and are very seedy, they need to be disposed of so that the seeds are not dispersed again into the garden.
- If the weeds have not yet developed seeds, they can be put into a compost bin.
- Students will complete their *Go Away Weeds* student investigation guide.

EVALUATION:

- Students will turn in their *Go Away Weeds* student investigation guide.
- Teacher will observe reactions and interactions with other students and the weeds.

GO AWAY WEEDS

Name

____

What is a weed?

____

____

Weeds compete with garden plants for which resources?

____

____

Sketch a picture of your garden area before you removed the weeds.

THIRD GRADE LESSON PLAN
Measuring the Garden

SUBJECT / GRADE LEVEL: Third Grade - Math.

ACADEMIC STANDARDS AND PERFORMANCE INDICATORS:

Math:

3.MDA.2 Estimate and measure liquid volumes capacity in customary units i.e., c., pt., qt., gal. and metric units mL, L to the nearest whole unit.

LESSON OBJECTIVES:

TSW explore what can be measured in the garden, determine what units and tools they need to make measurements, form estimations, and make exact measurements. TSW also explore what kind of measurements gardeners and farmers use.

MATERIALS:

- Measurement chart for each student (attached)
- Clip boards
- Chart paper or board for the teacher to record a class measurement chart
- Pencil for each student or student team
- Measuring tools
- *Measuring the Garden* student investigation guide (attached)
- *Measuring the Garden* teacher guide (attached)

ENGAGEMENT:

- Teacher will pose the question, "**What can be measured in the garden?**"

EXPLORATION:

- Teacher will lead students through the garden to locate items for the measurement exercise.

EXPLANATION:

- Class will record measurements on classroom sheet.
- Students will complete *Measuring the Garden* student investigation guide.

ELABORATION:

- Students can test their measuring skills on other garden items throughout the school year an example would be weighing produce as it is harvested.

EVALUATION:

- Students will turn in the *Measuring the Garden* student investigation guide.
- Teacher will observe reactions and interactions with other students and the measuring devices.

RESOURCES:

1. *Measuring the Garden* handout (attached).

TEACHER GUIDE

MEASURING THE GARDEN

BACKGROUND:

Students should understand that there are many items in a garden that can be measured. There are many measurement tools that can assist with each task.

ENGAGEMENT:

- Teacher will pose the question, "**What can be measured in the garden?**"

EXPLORATION:

- Allow students to walk through the garden with a partner or group to aid in generating ideas. Students will revert back to the measuring basics suggesting measuring length. This is great, but encourage them to continue brainstorming and discussing what other measurements could be made.
- Bring the class back together to share ideas. Make a list of all measurements students would like to make.
- Possible student ideas:
  –Perimeter of a garden bed
  –Circumference
  –Temperature in the sun
  –Temperature in the shade
  –Rainfall
  –Area of the garden or section
  –Weight of a planter
  –Weight of a wheel barrow
  –Wind direction or speed
  –The amount of time a plant is in direct sunlight in a day
  –The height of a sunflower
  –The length of the longest and shortest bean pod
- After students have recorded what the class is going to measure, they must determine what units they will use to make their measurements. Provide students time to discuss this with their partner or group before adding to the class chart. Students may need to reference unit anchor charts.

- Units may include inches, centimeters, feet, yards, meters, square units, ounces, pounds, grams, kilograms, minutes, hours, Celsius, Fahrenheit, etc.
- Next, allow students time to determine what tools they will need and add to the class chart. Tools may include thermometers, meter/yard sticks, scales, measuring tape, rain gauge, clocks, etc...
- Collect tools to measure.
- Adjustments may need to be made to the units selected based on the availability of measuring tools. Also, some measurements may require more than one day for an exact measurement.
- Before making exact measurements, students should estimate units. Students will need to examine what is being measured to make close estimations. Students may also be guided to consider using non-standard units and comparisons when making estimations.
- Example prompt questions:
  – **How many pounds might a sweet potato weigh if we know that one pound is about as heavy as a loaf of bread?**
  – **If you weigh 87 pounds, about how many pounds would you estimate the shovel weighs?**
  – **If a centimeter is about the length of your pinky nail, about how many centimeters long will the object be?**
- Next, students will use the measuring tools collected to make exact measurements. Ask students to consider if their answers are reasonable and encourage students to check their partner s or group s accuracy.
  ****It may be helpful to number the measurements by stations so that each student/group will receive the same experience.*

EXPLANATION:

- Once the class has been reassembled, record their measurements on the class chart and instruct students to record what they notice about the data (true statements that reflect the data). For example, a student may suggest that a planter is lighter than a wheel barrow, or that the height of a sunflower in inches is equivalent to 2 yards.
- If a group presents a measurement that does not seem accurate, ask students to determine why the measurement is inaccurate.
- Finally, ask students to discuss when a gardener or farmer may need to make measurements and allow time for them to share their reasoning. Students may also research garden measurements in the classroom.
- Students will complete *Measuring the Garden* student investigation guide.

ELABORATION:

- Measure the height of a group of plants or weight / length of harvested fruits / vegetables, graph data and determine the mean, median, and mode.
- Ask students to create a test garden measurement problem for their classmates to solve.

EVALUATION:

- Students will turn in the *Measuring the Garden* student investigation guide.
- Teacher will observe reactions and interactions with other students and the measuring devices.

MEASURING THE GARDEN

Name

____

Sketch four objects that you measured in the garden.

Object 1	**Object 2**
Object 3	**Object 4**

Complete the measurement for each object.

	Measurement
Object 1	
Object 2	
Object 3	
Object 4	

THIRD GRADE LESSON PLAN
Time to Transplant

SUBJECT / GRADE LEVEL: Third Grade - Life Science.

ACADEMIC STANDARDS AND PERFORMANCE INDICATORS:

Science:

3.L.5B.1 Obtain and communicate information to explain how changes in habitats such as those that occur naturally of those caused by organisms can be beneficial or harmful to the organisms that live there.

3.L.5B.2 Develop and use models to explain how changes in a habitat cause plants and animals to respond in different ways.

Math:

3.MDA.5 Understand the concept of area measurement. a. Recognize area as an attribute of plane figures; b. Measure area by building arrays and counting standard unit squares; c. Determine the area of a rectilinear polygon and relate to multiplication and addition.

LESSON OBJECTIVES:

TSW estimate the space needed to allow seedlings to grow into adult plants. TSW formulate a plan for transplanting seedlings into available garden space using measurement.

MATERIALS:

- Seedlings
- Measuring tape
- Information on size of adult plant for your seeds
- Fertilizer compost tea, fish emulsion, manure, etc.
- Classroom space to model available planting space
- Trowels
- *Time to Transplant* student investigation guide (attached)
- *Time to Transplant* teacher guide (attached)

ENGAGEMENT:

- Teacher will pose the question, "**What do the plants need in order to continue to thrive and grow?**"

EXPLORATION:

- Teacher will pass out seedlings to student groups / pairs.
- Students will think-pair-share what they think the plants need in order to continue to grow.

EXPLANATION:

- Students should plant their seedlings and fertilize half of the seedlings this will start the experiment in which students measure the seedlings for twenty days and graph the results.

ELABORATION:

- Students should regularly check on their seedlings 2 or 3 times a week time should be given for qualitative and quantitative observations.
- Students will complete *Time to Transplant* student investigation guide.

EVALUATION:

- Students will turn in the *Time to Transplant* student investigation guide.
- Teacher will observe reactions and interactions with other students and the plants.

RESOURCES:

1. *Time to Transplant* handout (attached).

TEACHER GUIDE
TIME TO TRANSPLANT

BACKGROUND:

Students should understand transplanting seedlings is an important step in growing a successful garden.

ENGAGEMENT:

- Teacher will pose the question, "**What do the plants need in order to continue to thrive and grow?**"

EXPLORATION:

- Pass out seedlings to student groups / pairs. Students think-pair-share what they think the plants need in order to continue to grow.
- Discuss as a class what the students came up with. If not said, the teacher needs to suggest that the plants be transplanted outside. Lead a discussion as to why they need to be moved outside.
- Make a T-Chart on the board labeled "Inside Conditions / Outside Conditions" and have the students list the differences. Go back and as a class, circle the differences that would affect a plant's life and viability. If not listed make sure to add wind, amount of sunlight / shade, water, temperature, animal interaction, competition over space / soil nutrients, etc...

EXPLANATION:

- **How much space will we need for our seedlings?** Share with students, or have students research, the information on the size and space needed for our seedlings once they are mature plants. Discuss the suggested spacing between plants, root depth, amount of sunlight, etc.
- Discuss ways to ensure that we provide the plants with this criteria.
- Optional: Students should make circles to represent the amount of space needed for the adult plants example: a plant needing 18" between would need a circumference of 36". These can be taken out to the garden and placed in the beds by the students to ensure that adequate space is provided. Make sure students are involved in problem solving and brainstorming ways to ensure that plants are provided enough space.
- Students take measuring tapes if previous optional activity is not done and measure spacing between plants to ensure they have enough space to grow.

- Take this time to review what living things need in order to survive. **What would happen if we didn t give the seedlings enough space? Food? Sunlight?**
- Explain that we are going to fertilize, or feed, only half of our seedlings. **Ask students to predict what they think will happen to the unfertilized plants? What about the fertilized plants?**
- Students should plant their seedlings and fertilize half. All of the seedlings should be watered.

ELABORATION:

- Students should regularly check on their seedlings 2 or 3 times a week. Time should be given for qualitative and quantitative observations.
- Students will complete *Time to Transplant* student investigation guide.

EVALUATION:

- Students will turn in the *Time to Transplant* student investigation guide.
- Teacher will observe reactions and interactions with other students and the plants.

TIME TO TRANSPLANT

Name ____

Chart the growth of your transplants.

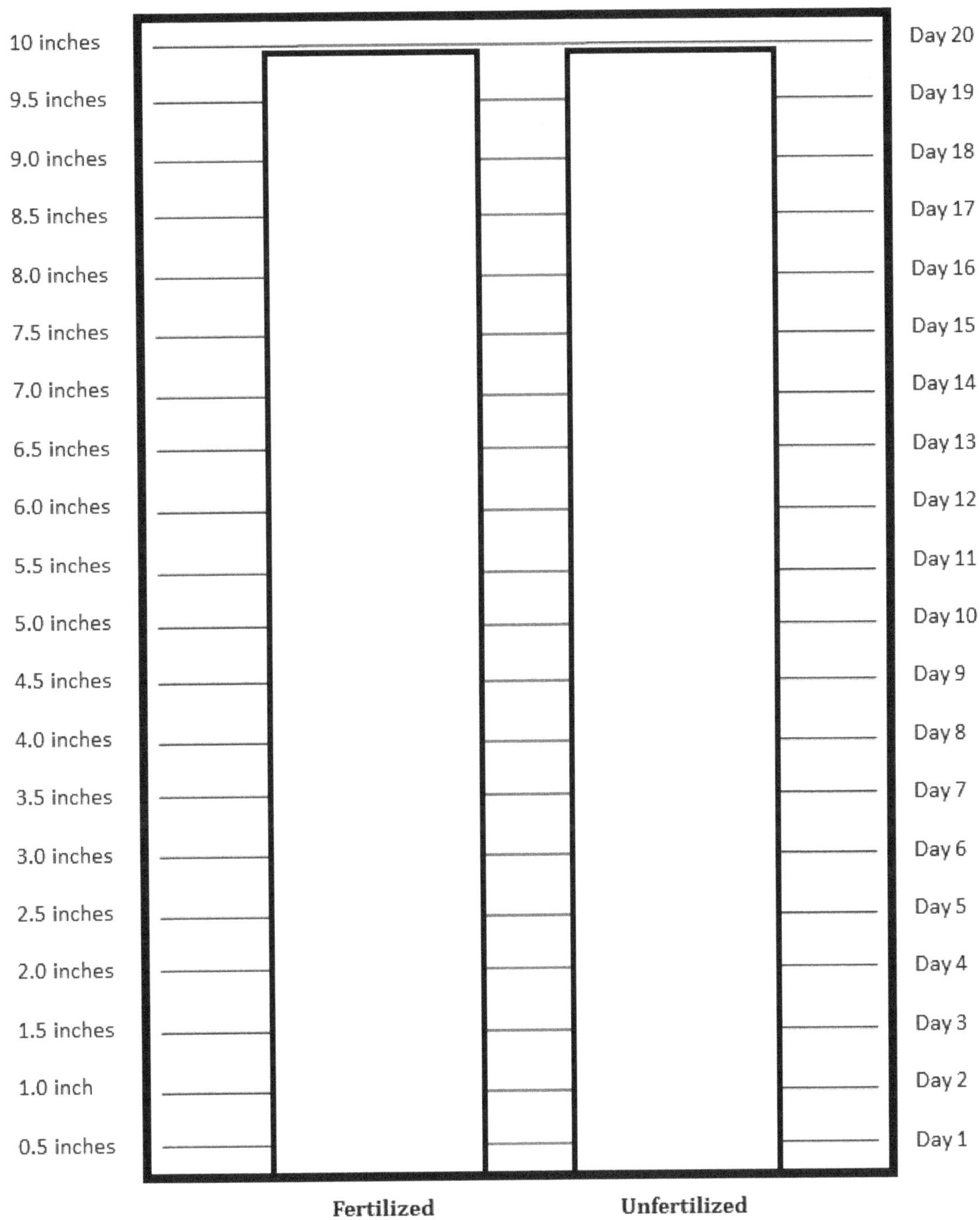

THIRD GRADE LESSON PLAN
Final Scavenger Hunt

SUBJECT / GRADE LEVEL: Third Grade - Inquiry and Life Science.

ACADEMIC STANDARDS AND PERFORMANCE INDICATORS:

Science:

3.L.5A.2 Develop and use a food chain model to classify organisms as producers, consumers, and decomposers and to describe how organisms obtain energy.

3.S.1A.1 Ask questions that can be 1.) answered using scientific investigations or 2.) used to refine models, explanations, or designs.

3.S.1A.3 Plan and conduct scientific investigations to answer questions, test predictions and develop explanations: 1.) formulate scientific questions and predict possible outcomes, 2.) identify materials, procedures, and variables, 3.) select and use appropriate tools or instruments to collect qualitative and quantitative data, and 4.) record and represent data in an appropriate form. Use appropriate safety procedures.

3.S.1A.4 Analyze and interpret data from observations, measurements, or investigations to understand patterns and meanings.

3.S.1A.6 Construct explanations of phenomena using 1.) scientific evidence and models, 2.) conclusions from scientific investigations, 3.) predictions based on observations and measurements, or 4.) data communicated in graphs, tables, or diagrams.

3.S.1A.7 Construct scientific arguments to support claims, explanations, or designs using evidence from observations, data, or informational texts.

3.L.5B.1 Obtain and communicate information to explain how change s in habitats such as those that occur naturally of those caused by organisms can be beneficial or harmful to the organisms that live there.

3.L.5B.2 Develop and use models to explain how changes in a habitat cause plants and animals to respond in different ways.

3.E.4A.3 Obtain and communicate information to exemplify how humans obtain, use, and protect renewable and nonrenewable Earth resources.

3.E.4B.4 Define problems caused by a natural event or human activity and design devices or solutions to reduce the impact on the environment.

3.L.5B.1 Obtain and communicate information to explain how changes in habitats such as those that occur naturally or those caused by organisms can be beneficial or harmful to the organisms that live there.

3.S.1B.1 Construct devices or design solutions to solve specific problems or needs: 1.) ask questions to identify problems or needs, 2.) ask questions about the criteria and constraints of the devices or solutions, 3.) generate and communicate ideas for possible devices or solutions, 4.) build and test devices or solutions, 5.) determine if the devices or solutions solved the problem and refine the design if needed, and 6.) communicate the results.

3.E.4B.4 Define problems caused by a natural event or human activity and design devices or solutions to reduce the impact on the environment.

Math:

3.MDA.2 Estimate and measure liquid volumes capacity in customary units i.e., c., pt., qt., gal. and metric units mL, L to the nearest whole unit.

LESSON OBJECTIVE: TSW explain the process of decomposition and justify its importance in a food chain. TSW observe the decomposition process and analyze what is occurring. TSW construct reasonable explanations for how the decomposition process occurs. TSW research the insects that are beneficial and harmful to the garden. Students will design, test, and improve modify solutions to attract beneficial insects and to deter harmful insects from the garden. TSW will explore the resources that weeds can take from garden plants such as sunlight, water and nutrients. TSW identify and remove unwanted weeds from the garden. TSW explore what can be measured in the garden, determine what units and tools they need to make measurements, form estimations, and make exact measurements. TSW also explore what kind of measurements gardeners and farmers use. TSW closely observe soil samples from around the school to determine the components of soil. TSW discuss the definition, function and importance of pollination. TSW discuss pollinator habitats and the ways in which humans affect pollinator/plant relationships in negative and positive ways. TSW analyze data to predict how many pollinators will visit the garden. TSW explain the difference between renewable and nonrenewable resources; TSW identify examples of each. TSW justify the importance of reducing, reusing and recycling to help maintain Earth's supply of nonrenewable resources. TSW estimate the space needed to allow seedlings to grow into adult plants. TSW formulate a plan for transplanting seedlings into available garden space using measurement. TSW investigate the body parts of worms as well as their reactions to stimuli such as light, vinegar, shelter rocks in their habitat soil.

ENGAGEMENT:

- Teacher will pose the question, "**What is a term that you've learned during our garden lessons?**"

MATERIALS:

- *Disappearing Banana Peel* student investigation guide
- *Engineering in the Garden* student investigation guide
- *Go Away Weeds* student investigation guide
- *Measuring the Garden* student investigation guide
- *Pollinator Habitats* student investigation guide
- *Renewable and Nonrenewable Resources* student investigation guide
- *Time to Transplant* student investigation guide
- *Worms at Work* student investigation guide
- *Peek Inside the Soil* student investigation guide
- Paper lunch sacks to collect items

EXPLORATION:

- Teacher will poll students for plant terms until all of the terms for the scavenger hunt have been discussed.

EXPLANATION:

- Using the *Scavenger Hunt* chart, the teacher will guide students through the explanation of all terms in the chart.
- After reviewing the *Scavenger Hunt* chart, students will go to the garden in order to find the items.

ELABORATION: IN THE CLASSROOM

- Students can compost all of the items from the scavenger hunt including the chart and the paper bag.
- If there is nowhere to compost the items the activity could also be turned into a digital scavenger hunt, where students simply take pictures of the items in the chart.

EVALUATION: IN THE CLASSROOM

- Students will show the teacher items in their bags so that the teacher may check them off during the Scavenger Hunt.

RESOURCES:

1. *Scavenger Hunt* handout (attached).

SCAVENGER HUNT

Name ____

Instructions: Find all of the items listed below:

Item	Where did you find it?
1) Plant visited by beneficial insects	
2) A weed	
3) An item measured in the garden	
4) Soil sample from the garden	
5) Soil sample from other area	
6) One source of pollen	
7) One source of nectar	
8) Renewable resource	
9) Nonrenewable resource	
10) Item from a worm habitat	

LESSONS FOR FOURTH GRADE

FOURTH GRADE LESSON PLAN
Plant Life Cycles

SUBJECT / GRADE LEVEL: Life Science - 4th Grade - Plant Life Cycles.

ACADEMIC STANDARDS AND PERFORMANCE INDICATORS:

Science:

4.L.5A.2 Analyze and interpret data from observations and measurements to compare the stages of development of different seed plants.

LESSON OBJECTIVES:

Using observation and measurement, TSW compare the life cycle stages of 3 different seed plants.

MATERIALS:

- 3 garden spaces ready for planting
- 3 different types of seeds must match growing conditions for the region time of year
- 3 small pots
- Popsicle sticks or wooden stakes to label plants
- Plastic bags or plastic containers with lids
- Paper towels
- *Plant Life Cycle* student investigation guide (attached)
- *Plant Life Cycle* teacher guide (attached)

ENGAGEMENT:

- Teacher will ask students to describe the stages of a human's life cycle infancy, adolescence, adulthood.
- Point out that even though all humans go through the infancy stage, they mature or develop at different rates. Examples are weight, height, learning to walk, learning to talk, etc... Likewise, humans have the same anatomy but different characteristics. Examples are hair, eyes, and skin color.

EXPLORATION:

- Students will be making observations and measurements for many weeks to compare the life cycle stages of three different seed plants. One set will be grown in the garden while the other will be grown indoors in pots.

EXPLANATION:

- Using the *Plant Life Cycles* teacher guide, the teacher will guide students through the completion of the *Plant Life Cycles* student investigation guide.

ELABORATION:

- Students will complete *Plant Life Cycles* student investigation guide.

EVALUATION:

- Students will turn in the *Plant Life Cycles* student investigation guide.
- Teacher will observe reactions and interactions with other students and the plants.

RESOURCES:

1. *Plant Life Cycles* handout (attached).

TEACHER GUIDE
PLANT LIFE CYCLES

BACKGROUND:

Students should understand that plants have three different life cycles that take place between seed germination and death.

ENGAGEMENT:

- Teacher will ask students to **describe the stages of a human's life cycle** infancy, adolescence, adulthood.
- Point out that even though all humans go through the infancy stage, they mature or develop at different rates. Examples are weight, height, learning to walk, learning to talk, etc. Likewise, humans have the same anatomy but different characteristics. Examples are hair, eyes, and skin color.

EXPLORATION:

Students will be making observations and measurements for many weeks to compare the life cycle stages of three different seed plants. Explain that all three plants may go through the stages of dormancy, germination, seedling, mature plant, and death. Students will investigate the similarities and differences of each plant at each stage for several weeks.

- Guide students through following the planting instructions on seed packets for how much space each plant will need and how deep to plant the seeds in the soil.
- Plant the three different types of seeds in three different rows or plots. ****Plant only one type of seed in each row or plot. Water the seeds.*
- Mark each row with some type of label to remember what was planted.
- Take some of each seed type back to the classroom.
- In order for students to closely observe germination of the seeds, germinate the unplanted seeds in the classroom.
- For each group, add seeds to a plastic container with a lid or a plastic bag. Add a paper towel to the bag or container. Ask students if they have an idea about what seeds need to germinate.
- A common misconception is that seeds need sunlight to germinate. Seeds only need water and warmth. If sunlight is suggested, place the seeds in a dark cabinet to illustrate the point. Students may suggest that seeds need soil to germinate. Explain that the seeds have enough food

inside of them cotyledon to germinate. Once students suggest adding water, add water and place the seeds in a cabinet or on a counter away from the window.

- Allow students to observe the germinating seeds.
- The roots will grow first, followed by the stem, and emergence of leaves.
- Plant the different seeds in different clay pots and keep in the classroom. Allow students to continue observing when the plants become seedlings, mature plants, and finally die.
- Potted seedlings need to be placed in the window to use sunlight for photosynthesis.
- There are a few reasons to plant the three seed types in the classroom:
  1. Students can closely observe each stage of the plant s life cycles.
  2. Students will see that the plants in the classroom will not grow as well as the plants in the garden and may develop theories for why this is happening. Students will notice that the plants in the garden have more space. They will actually be able to observe the roots growing out of the bottom of the small pots.
  3. The pots can be taken outside and placed next to the plants growing in the garden to make close comparisons.

EXPLANATION:

Guide students through the completion of their *Plant Life Cycles* student investigation guide using the information below:

- Seed: When pollination occurs, seeds are produced and may be stored in fruits. Seeds are covered by a protective layer called a seed coat. Seed coats allow seeds to remain "asleep" or dormant until conditions are right for germination. Seeds contain tiny undeveloped plants embryo and a little lunch box of food cotyledon for growth to start. Most seeds need water and warmth to germinate begin to grow.
- Seedling: Seedlings are baby plants that will grow into mature adult plants. Seedlings first produce roots which begin to grow and take in nutrients and water from the habitat. The stem then starts to grow towards light and the first leaves form on the stem. Later, more leaves will form that help the plant make its food.
- Mature Plant: Mature plants have the same structures for example roots, stems, and leaves as seedlings, but in addition they develop flowers or cones, which produce seeds. A plant s life cycle can be described by how long a plant lives or how long it takes to grow, flower, and produce seed.
- Annual: An annual is a plant that completes its life cycle in one growing season. It will grow, flower, produce seed, and die.
- Perennial: A perennial is a plant that lives for 3 or more years. It can grow, flower, and produce seed for many years before dying.

- *Biennial*: A biennial is a plant that needs two growing seasons to complete its life cycle. It produces leaves in the first season. Then it goes dormant or rests over the winter. In the spring, it will begin to grow again and grow flowers, produce seed, and die. The seed that is left behind on the ground germinates and the cycle begins again.
- Vocabulary: *germination*, *seed*, *seedling*, *mature adult plant*, *reproduction*, *seed coat*, *cotyledon*, *embryo*, *dormant*.

ELABORATION:

- Over the course of many weeks, make frequent trips to the garden. Students will take measurements of the three different plants at different stages. While they won't be measuring every plant, guide students to measure one of the plants from each row every time they take measurements. Students may start out by choosing to measure the tallest seedling from each row. Whatever is decided, make sure to return to the same plant for each measurement. Mark the plant chosen with a Popsicle stick.
- Guide students to take many measurements that to compare plant growth. They could be height, number of leaves, leaf appearance, leaf width, circumference of the stem, the amount of time it takes for the plant mature develop flowers , etc. Students should record data on an individual or classroom sheet.
- Students will complete *Plant Life Cycles* student investigation guide.

EVALUATION:

- Students will turn in the *Plant Life Cycles* student investigation guide.
- Teacher will observe reactions and interactions with other students and the plants.

PLANT LIFE CYCLES

Name

____

What is difference between a seedling and a mature plant?

____

____

____

____

What is germination?

____

____

____

____

List the three plant life cycles.

Plant Life Cycle

____

Plant Life Cycle

____

Plant Life Cycle

____

FOURTH GRADE 5E LESSON PLAN
Catching Air

SUBJECT / GRADE LEVEL: Fourth Grade - Earth Science - Weather Station Creation Part 1.

ACADEMIC STANDARDS AND PERFORMANCE INDICATORS:

Science:

4.E.2B.1 Analyze and interpret data from observations, measurements, and weather maps to describe patterns in local weather conditions including temperature, precipitation, wind speed/direction, relative humidity, and cloud types and predict changes in weather over time.

4.S.1A.4 Analyze and interpret data from informational texts, observations, measurements, or investigations using a range of methods such as tabulation or graphing to 1.) reveal patterns and construct meaning or 2.) support explanations, claims, or designs.

4.S.1B.1 Construct devices or design solutions to solve specific problems or needs: 1.) ask questions to identify problems or needs, 2.) ask questions about the criteria and constraints of the devices or solutions, 3.) generate and communicate ideas for possible devices or solutions, 4.) build and test devices or solutions, 5.) determine if the devices or solutions solved the problem and refine the design if needed, and 6.) communicate the results.

MATERIALS:

- Plastic straws
- Thin ribbon or yarn 8 inches in length
- Clipboard
- *Catching Air Data Collection Table* attached)
- *Scientific Instruments/Technology Chart* (attached)
- Tape

For each pair:

- Clear plastic cup
- Bubble solution and wand

For the class:

- Magnetic compass
- Anemometer
- PVC pipe frame
- Rope
- Wind ties
- *Beaufort Scale* http://www.crh.noaa.gov/images/iwx/publications/Beaufort_Wind_Chart.pdf
- *Catching Air* student investigation guide (attached)
- *Catching Air* teacher guide (attached)

LESSON OBJECTIVES:

TSW construct weather instruments to measure wind direction and wind speed and determine the helpful and harmful ways wind affects the garden.

ENGAGEMENT: IN THE CLASSROOM

- Teacher will pose the question, "**Is wind helpful or harmful to plants?**"

EXPLORATION: IN THE GARDEN

- Students will construct a wind indicator tool using straws and ribbon / yarn.

EXPLANATION: IN THE GARDEN AND CLASSROOM

- Using the *Catching Air* teacher guide, the teacher will guide students through the completion of the *Catching Air* student investigation guide.

ELABORATION:

- Students will complete *Catching Air* student investigation guide.

EVALUATION:

- Students will turn in the *Catching Air* student investigation guide.
- Teacher will observe reactions and interactions with other students and the wind indicator tool.

RESOURCES:

1. *Beaufort Wind Chart*: http://www.crh.noaa.gov/images/iwx/publications/Beaufort_WindChart.pdf.
2. *Catching Air Data Collection* table (attached).
3. *Scientific Instruments/Technology* chart (attached).
4. *Catching Air* handout (attached).

TEACHER GUIDE
CATCHING AIR

BACKGROUND:

Students should understand that wind direction and speed can affect plant growth in the garden.

ENGAGEMENT:

- Teacher will pose the question, "**Is wind helpful or harmful to plants?**"

EXPLORATION:

This lesson is Day 1 of a 6-day teaching / learning sequence. Data-collection is on-going throughout the year. It is important to select a start day with at least slightly windy conditions.

- **What causes wind?** Small groups discuss and share responses as a class.
- **How do we measure wind speed and wind direction? How does wind affect our garden in helpful and harmful ways?**
- Distribute straws and ribbon / yarn and explain to students that they will make a wind indicator tool. Ask, "**How will these materials help us understand wind?**"
- *Wind direction* is the measure of where the wind is coming from, not where it is headed.

EXPLANATION:

- Determine the direction of the wind. Practice describing the wind's direction orally and in writing. Describe the speed of the wind using adjectives: still, light, breezy, gusty, steady, strong, hazardous, etc...
- Wind Direction Indicator tool: Instruct students to place a piece of yarn or ribbon half way down a plastic straw. Secure the ribbon or yarn to the straw with a piece of tape. Hold it above your head to determine which way it moves. Allow time for students to use the straw and ribbon / yarn to discover the direction of the wind. Discuss and record the factors that influence the effectiveness of the device (i.e. obstructions) and considerations used during design (i.e. length of string). COOL TRICK: Demonstrate how to find the wind s direction by slowly turning your head through the wind from shoulder to shoulder. When the wind blows on both ears at the same time, your nose will be pointing into the wind Line the class up or spread them out in a manner that ensures they are not obstruction each other.
- Catching Air: Form the class into face-to-face partner lines so each student faces a partner. Give students on one side a clear plastic cup and students on the other side a bubble solution and

wand. Challenge students to blow bubbles into their partner's cup using only the wind speed and direction. Allow students to move to any area with enough space for them to have success. Switch roles and repeat. Discuss results.

- Ask students to think of all the ways, both helpful and harmful, that the wind affects our garden.
- Show students the PVC pipe structure, rope, wind ties, anemometer, wind vane, and magnetic compass and explain that this frame is the foundation for our garden's Weather Station.
- Students draw the frame, including labeled measurements of its dimensions, in their science notebooks and indicate where they think the anemometer, wind vane, and compass should attach. Use the provided materials and student reasoning to attach the devices. Discuss. Return to the classroom.
- Students begin a *Scientific Instruments / Technology Chart* in their science notebooks (attached) to sketch, define, and explain the problem it solves.
- **How does wind affect our garden?** Discuss the ideas students came up with in the garden and categorize the effects on a T-Chart labeled harmful and helpful. Guide students with questions to list several examples of helpful (carries scents of flowers, spreads seeds, carries insects, etc.) and harmful (erosion of topsoil, uproots plants, breaks stems, etc.) effects.
- Brainstorm ways to prevent harmful effects of the wind and use helpful effects of the wind to benefit the garden.
- Vocabulary: *technology, equipment, instruments, meteorology, wind speed, wind direction, compass, anemometer, wind vane sock , digital, analog, weather vane, compass rose.*

ELABORATION:

- *CATCHING AIR* DATA COLLECTION: Students record data about the wind throughout the school year (attached). They use the wind indicators from the garden's Weather Station to determine the direction of the wind. They write qualitative descriptions of the wind speed. They conduct online research (http://www.weather.gov/chs/) to determine the maximum and minimum wind speed for the day quantitative and calculate the range. Students record this data in a section of their science notebooks.
- Beaufort Scale: http://www.crh.noaa.gov/images/iwx/publications/Beaufort_Wind_Chart.pdf.
- Have students analyze and interpret the Beaufort Scale and discuss implications for the garden.
- **At which wind speeds do we need to think about the health of the garden?**
- Students will complete the *Catching Air* student investigation guide.

EVALUATION:

- Students will turn in the *Catching Air* student investigation guide.
- Teacher will observe reactions and interactions with other students and the wind indicator tool.

Catching Air Data Collection

Date/Time	Wind Direction	Wind Speed (Qualitative)	Wind Speed (Quantitative)		
			Maximum	Minimum	Range

Scientific Instruments/ Technology

Topic:

Device	Sketch	Definition	Problem it Solves
Magnetic Compass			
Anemometer			
Wind Vane			
Thermometer			
Hygrometer			
Barometer			

CATCHING AIR

Name

____

What does wind affect in the garden?

____

____

____

____

What causes wind?

____

____

____

____

How do we measure wind speed and wind direction?

____

____

____

FOURTH GRADE 5E LESSON PLAN
The Garden Is So Hot Right Now

SUBJECT / GRADE LEVEL: Fourth Grade - Earth Science - Weather Station Creation Part 2.

ACADEMIC STANDARDS AND PERFORMANCE INDICATORS:

Science:

4.E.2B.1 Analyze and interpret data from observations, measurements, and weather maps to describe patterns in local weather conditions including temperature, precipitation, wind speed/direction, relative humidity, and cloud types and predict changes in weather over time.

4.S.1A.3 Plan and conduct scientific investigations to answer questions, test predictions and develop explanations: 1.) formulate scientific questions and predict possible outcomes, 2.) identify materials, procedures, and variables, 3.) select and use appropriate tools or instruments to collect qualitative and quantitative data, and 4.) record and represent data in an appropriate form. Use appropriate safety procedures.

4.S.1A.4 Analyze and interpret data from informational texts, observations, measurements, or investigations using a range of methods such as tabulation or graphing to 1.) reveal patterns and construct meaning or 2.) support explanations, claims, or designs.

LESSON OBJECTIVES:

TSW measure soil temperatures to determine the best times to plant vegetable crops in the garden. TSW measure daily high and low air temperatures, find the range, and plot the data on a line graph. TSW determine the number of days in a given month there are for maximum plant growth, using thermo-period data. TSW compare air temperature in direct sunlight to air temperature in shade to make inferences about plant growth.

ENGAGEMENT:

- Teacher will pose the question, "**Why is temperature important to plants in the garden?**"

EXPLORATION:

- Students will investigate the relationship between temperature and plant growth this includes both soil and air temperature.

MATERIALS:

For each student

- Science notebooks
- Clipboard

For the class

- Graph paper
- Chart paper
- Colored pencils blue and red
- Ruler or meter stick
- Sticky notes
- Maximum/minimum thermometer probe thermometer
- Water thermometer

For each group

- Student thermometer
- Soil thermometer
- Screwdriver
- Soil Temperature Guidelines Table (attached)
- Calculator
- Temperature Reading Template (attached)
- *The Garden Is So Hot Right Now* student investigation guide (attached)
- *The Garden Is So Hot Right Now* teacher guide (attached)
- *Plant Life Cycle* teacher guide (attached)

EXPLANATION:

- Using *The Garden Is So Hot Right Now* teacher guide, the teacher will guide students through the completion of *The Garden Is So Hot Right Now* student investigation guide.

ELABORATION:

- Students will complete *The Garden Is So Hot Right Now* student investigation guide.

EVALUATION: IN THE CLASSROOM

- Students will turn in the *The Garden Is So Hot Right Now* student investigation guide.
- Teacher will observe reactions and interactions with other students and the thermometer.

RESOURCES:

1. Local Weather Information: www.wunderground.com.
2. *Soil Temperature Guidelines Table* (attached).
3. *Temperature Reading Template* (attached).
4. *Plant Life Cycles* handout (attached).

TEACHER GUIDE

THE GARDEN IS SO HOT RIGHT NOW

BACKGROUND:

Students should understand that plant growth is affected by multiple temperatures including air and soil temperature.

TEACHER PREPARATION:

Acquire materials one maximum/minimum thermometer for the class and one soil thermometer for each group. Maximum/minimum thermometers can be purchased for less than $20 at a local garden center. Local garden centers also sell simple soil thermometers for a few dollars each. They have a glass bulb and a strong metal tip. If you cannot access soil thermometers, then any thermometer will do as long as they measure temperatures down to freezing. Prepare to teach students the formula for finding the average mean of a data set.

ENGAGEMENT:

- Teacher will pose the question, "**Why is temperature important to plants in the garden?**"

EXPLORATION:

- Temperature is one of the environmental factors that affect plant growth. Along with light, water and fertility, temperature readings give gardeners information about when to plant, harvest, and protect plants. Daily temperature change, or thermo-period, is a key component of the relationship between temperature and plant growth. When day temperatures are 10^{0} to 15^{0} F (5.5^{0} to 8^{0} C) warmer than night temperatures, plants produce maximum growth.
- This lesson serves to establish a procedure for ongoing record-keeping and data analysis about daily temperature ranges. Each day, students plot the high and low temperature for a specific area in the garden on a line graph. They calculate this range daily and use the data to make decisions about planting and watering in the garden.
- Soil temperature is often just as important to plant growth as air temperature, so this lesson also introduces methods to monitor soil temperature in different areas of the garden. Seeds for different plants require different soil temperatures to properly germinate, so monitoring the soil temperature in the garden allows students to understand when the best times are to plant which crops. Establish a procedure for daily temperature readings and graphing by selecting appropriate times to measure temperature and plot the data on a line graph.

- **What is temperature?**
- Have students discuss these questions in small groups. Tell students that the air temperature today is ______ ^{0} F, according to www.wunderground.com.
- Ask, "**Do you think that is the actual temperature in our garden?**" Explain reasoning in small groups and share responses. "**Besides air temperature, what other temperatures are important in the garden?**" Guide students to soil temperature, water temperature, direct sun temperature, and shade temperature.
- Focus questions:
  –What information does soil temperature give gardeners?
  –What is the range of daily high and low air temperature in the garden?
  –How many other temperatures can we discover in the garden?
- Prepare the class for outdoor activities while students format their science notebooks with heading, date, time, and focus questions.
- Place students in groups of three or four students per group. Demonstrate how to make a pilot hole for measuring soil temperature. When using a thermometer to measure soil, make a pilot hole with a screwdriver so you don't damage the thermometer. Take your soil thermometer readings at least 5–6 inches deep in the soil so that the soil temperature is being measured, rather than the air.
- Explain the task to students: You have about ten minutes to measure and record as many different temperatures as you can find in the garden. Make sure that you record the temperatures in Fahrenheit, the exact location of your measurement, and which temperature you recorded air, direct sun, shade, water, soil in sweet potato beds, etc. Groups should measure at least five different temperatures. Once you record all of your measurements, types, and locations in your science notebooks, think about the factors responsible for the different temperatures and list them in your science notebooks.
- On chart paper, prepare a T-Chart on which students record their temperature readings, their locations, and the types of temperatures they found.
- Ask, "**What do you notice about the data in our chart?**" Discuss student responses as a class and begin listing the factors responsible for the different temperatures below the chart.

EXPLANATION:

- Give each group a *Soil Temperature Guidelines Table* (attached) and give them a few minutes to read and interpret the information. Discuss what students notice. Ask students a series of questions to assess their understanding of the table. For example: **Which vegetables would be best to plant today? When would be the best time to plant kale? When would be the best time to plant tomatoes?**

- Show students the maximum / minimum thermometer and have them predict its function / purpose. A maximum / minimum thermometer is an instrument used to measure the daily high and low temperatures for a particular area. **What time of day is the temperature the coolest? What time of day is the temperature the warmest? Why?** Recall how to find the range of daily temperatures and record the equation in science notebooks. Explain a procedure for measuring and recording daily high and low temperatures, and students create a data table in their science notebooks to record these data for one week. Tell them that a *thermo-period* refers to the range of daily temperature and the ideal range from night to day temperatures for maximum plant growth is 10–15⁰ F. As we take daily temperature readings and find the range, we will always be looking for this optimal thermo-period, so we will know when our plants should experience the most growth. At the end of one week, students determine which days offer a temperature range ideal for plants and which days we should expect less growth.
- After recording the soil temperature for one week at the same time each day (mid-day), help students calculate the average soil temperature for the week. Have them record the formula for finding the average in their science notebook. They may also calculate the weekly average for high temperature, low temperature, and range of daily temperature.
- Attach a maximum / minimum thermometer and a soil thermometer to the Weather Station in the garden.
- Vocabulary: *temperature*, *thermometer*, *thermal*, *high*, *maximum*, *low*, *minimum*, *range*, *shade*, *direct sunlight*, *optimal*, *Fahrenheit*, *Celsius*, *soil*, *air*, *water*, *degrees*, *thermo-period*, *average*, *mean*.

ELABORATION:

- Students develop an open-ended investigation about temperature in the garden independently or in pairs. They may a test how temperature affects a) specific type of plant, b) compare two specific plants and test how they respond to changes in temperature, or c) compare soil temperature to air temperature over a given amount of time. Have students develop a hypothesis, a plan, and an independent investigation.
- After several weeks of graphing high / low temperatures and soil temperatures, students use data to inform gardening decisions. They begin making an ongoing list of suggestions for the garden based on temperature data.
- Always encourage students to make predictions for each month. Likewise, have them record their "wonderings" on sticky notes for display to guide future investigations.
- Students graph the daily high and low temperatures in the garden on graph paper throughout the year, and teacher monitors the graph for accuracy. They do the same for soil temperatures.

- They use historical climate data to plot average daily temperature or record temperatures on the graph. They indicate the ideal thermo-periods on the daily high and low temperature graph and the ideal planting temperatures on the soil temperature graph.
- Students will complete the *The Garden Is So Hot Right Now* student investigation guide.

EVALUATION:

- Students will turn in the *The Garden Is So Hot Right Now* student investigation guide.
- Teacher will observe reactions and interactions with other students and the thermometer.

TEMPERATURE READING TEMPLATE

Weekly Soil Temperature

Date	Temperature
	Average

Weekly Air Temperature (Garden)

Date	High Temperature	Low Temperature	Range
Average			

Weekly Air Temperature (www.wunderground.com)

Date	High Temperature	Low Temperature	Range
Average			

SOIL TEMPERATURE GUIDELINES TABLE

SOIL TEMPERATURE	VEGETABLE CROPS
Crops that germinate down to 40° F (4° C)	lettuce, kale, peas, spinach, arugula, parsnips, bok choi, radishes, fava beans
Crops that germinate down to 50° F (10° C)	onions, leeks, turnips, Swiss chard
Crops that germinate down to 60° F (16° C)	broccoli, cabbage, carrots, beans, beets, cauliflower
Crops that germinate down to 70° F (21° C)	melons, corn, squash, cucumbers, peppers, eggplants, tomatoes

THE GARDEN IS SO HOT RIGHT NOW!

Name

____

How does soil temperature affect plant growth?

____

____

____

____

____

How does air temperature affect plant growth?

____

____

____

____

____

At what depth should soil temperature be taken and why?

____

____

____

____

FOURTH GRADE LESSON PLAN

Precipitation or Irrigation? Let's Take a Rain Check

SUBJECT / GRADE LEVEL: Fourth Grade - Earth Science - Weather Station Creation Part 3.

ACADEMIC STANDARDS AND PERFORMANCE INDICATORS:

Science:

4.E.2B.1 Analyze and interpret data from observations, measurements, and weather maps to describe patterns in local weather conditions including temperature, precipitation, wind speed/direction, relative humidity, and cloud types and predict changes in weather over time.

4.S.1A.4 Analyze and interpret data from informational texts, observations, measurements, or investigations using a range of methods such as tabulation or graphing to 1.) reveal patterns and construct meaning or 2.) support explanations, claims, or designs.

4.S.1B.1 Construct devices or design solutions to solve specific problems or needs: 1.) ask questions to identify problems or needs, 2.) ask questions about the criteria and constraints of the devices or solutions, 3.) generate and communicate ideas for possible devices or solutions, 4.) build and test devices or solutions, 5.) determine if the devices or solutions solved the problem and refine the design if needed, and 6.) communicate the results.

LESSON OBJECTIVES:

TSW use a rain gauge constructed from reusable materials to measure amounts of precipitation and monitor responsible irrigation practices using data collected from the homemade rain gauges.

ENGAGEMENT:

- Teacher will pose the question, "**Why should we measure the rain?**"

EXPLORATION:

- Teacher will introduce the rain gauge design and build challenge to the students.

EXPLANATION: IN THE GARDEN OR IN THE CLASSROOM

- Using the *Precipitation or Irrigation? Let's Take a Rain Check* teacher guide, the teacher will guide students through the completion of the *Precipitation or Irrigation? Let's Take a Rain Check* student investigation guide.

MATERIALS:

For the class:

- Various containers (recycled, reused, or repurposed)
- Tape
- Scissors
- Wind zip ties
- Graph paper for ongoing data collection and graphs
- Sticky notes

For each student:

- Science notebooks
- Graph paper
- Internet source to analyze average precipitation data

For each pair:

- Container
- Cm/mm ruler
- Permanent marker

Teacher Preparation:

- Gather containers ahead of time, such as glass jars or tubes, plastic jars or bottles, or coffee cans
- Create appropriate Data Collection Table
- *Precipitation or Irrigation? Let's Take a Rain Check* student investigation guide (attached)
- *Precipitation or Irrigation? Let's Take a Rain Check* teacher guide (attached)

ELABORATION:

- Students will continue to monitor rain gauge and collect weather data from the garden.

EVALUATION:

- Students will turn in the *Precipitation or Irrigation? Let's Take a Rain Check* student investigation guide.
- Teacher will observe reactions and interactions with other students and the rain gauge.

RESOURCES:

1. *Precipitation or Irrigation? Let's Take a Rain Check* handout (attached).

TEACHER GUIDE

PRECIPITATION OR IRRIGATION? LET'S TAKE A RAIN CHECK

BACKGROUND:

Students should understand that tracking precipitation is an important part of growing a successful garden.

ENGAGEMENT:

- Teacher will pose the question, "**Why should we measure the rain?**"

EXPLORATION:

- Introduce Challenge:
  –Today you will work with a partner to design and construct a device to measure precipitation in the garden. We will attach the most accurate and durable devices to our Weather Station.
- Focus Questions:
  –How can we measure precipitation in the garden?
  –How can we design and construct a rain gauge using reusable materials?
  –How can precipitation data collected from a rain gauge help us control the amount of water we use to irrigate the garden?
  –Why do gardeners and farmers need to control the amount of water distributed?
- Students work in pairs to develop designs and sketches for constructing a rain gauge to track the amount of precipitation for one month. Student pairs select materials of their choice and begin construction.
- Explain that the rain gauge must be durable enough to withstand rain and wind and must have a way to measure rainfall. Encourage students to use cm / mm to measure rainfall.
- Once student pairs complete their rain gauge construction, vote on which device we should attach to the Weather Station. Use wind zip ties to securely attach rain gauges to the Weather Station. Other student pairs select a location in which to attach their rain gauges.
- Next, students create a data-collection plan in their science notebooks time / amount of rain. They should consider comparing their data to average rainfall and official rainfall.

EXPLANATION:

- Once an air mass has too much moisture and cannot hold it anymore, the air mass has reached its saturation point. Once the saturation point is reached, precipitation occurs. Since precipitation is vital to plant growth, students should keep data on rainfall. The rule of thumb is an inch a week for garden plants. Devise a plan to record data on a monthly or weekly double bar graph to compare average precipitation with actual precipitation measured and official precipitation which can be found online.
- Rain gauges should be read as such: Once the amount of rainfall is recorded, the water in the rain gauge should be emptied so that accurate readings can be made in the future. If the water is left in the rain gauge for a long period of time after rainfall, the data becomes inaccurate because some of the water has evaporated.
- Allow students to answer focus questions independently in their science notebooks. Discuss student responses.
- Make sure they understand precipitation, rain gauge, data-collection methods, and implications for irrigation by asking appropriate follow-up questions. More rainfall requires less irrigation and less rainfall requires more irrigation.
- Establish procedures for collecting, comparing, and communicating monthly precipitation data on graph paper. After several months of modeling, gradually release this responsibility to students. After a measurable amount of rainfall, demonstrate how to calculate the average mean rainfall by dividing the sum of all precipitation levels by the total number of rain gauges.
- Write and explain the equation used to find the class average depending on how proficient the students are with operations, this lesson may require a separate session.
- Students will complete the *Precipitation or Irrigation? Let's Take a Rain Check* student investigation guide.
- Vocabulary: *precipitation, data, variables, independent variable, dependent variable, rain gauge, rain, sleet, snow, hail, measurements, precise, predict, irrigation, drizzle, ice pellets, meteorology, meteorologist, saturation point, cloud cover, stratus clouds, cumulonimbus clouds.*

ELABORATION:

- Ask the students: "**What variables affect the amount of precipitation collected in a given rain gauge?**" Discuss student ideas and as they talk about it, list the variables on chart paper and in science notebooks.
- Students work in pairs to compare their rain gauge with that of another pair and list the advantages and disadvantages of the design and / or construction methods.

- Ask, "**How do our measurements and calculations compare with official National Weather Service data?**" Demonstrate for students how to access historical climate data from a reputable source as a means of comparison.
- As students graph the rainfall they measure in the garden from month to month, have them create double-bar graphs to compare average rainfall to actual rainfall.
- Develop an "Irrigation Action Plan" with students to:
  1. Estimate amounts of harvested water in the cistern.
  2. Make irrigation decisions.
  3. Analyze affects on the garden.
- Always encourage students to make predictions for each month. Likewise, have them record their "wonderings" on sticky notes for display to guide future investigations.

EVALUATION:

- Students will turn in the *Precipitation or Irrigation? Let's Take a Rain Check* student investigation guide.
- Teacher will observe reactions and interactions with other students and the rain gauge.

PRECIPITATION OR IRRIGATION? LET'S TAKE A RAIN CHECK!

Name

____

Sketch your rain gauge design below.

List the materials you will need for construction.

____

____

____

____

FOURTH GRADE LESSON PLAN
Hygrometers, Psychrometers and Barometers

SUBJECT / GRADE LEVEL: Fourth Grade - Earth Science - Weather Station Creation Part 4.

ACADEMIC STANDARDS AND PERFORMANCE INDICATORS:

Science:

4.E.2B.1 Analyze and interpret data from observations, measurements, and weather maps to describe patterns in local weather conditions including temperature, precipitation, wind speed/direction, relative humidity, and cloud types and predict changes in weather over time.

4.S.1A.3 Plan and conduct scientific investigations to answer questions, test predictions and develop explanations: 1.) formulate scientific questions and predict possible outcomes, 2.) identify materials, procedures, and variables, 3.) select and use appropriate tools or instruments to collect qualitative and quantitative data, and 4.) record and represent data in an appropriate form. Use appropriate safety procedures.

4.S.1A.4 Analyze and interpret data from informational texts, observations, measurements, or investigations using a range of methods such as tabulation or graphing to 1.) reveal patterns and construct meaning or 2.) support explanations, claims, or designs.

4.S.1B.1 Construct devices or design solutions to solve specific problems or needs: 1.) ask questions to identify problems or needs, 2.) ask questions about the criteria and constraints of the devices or solutions, 3.) generate and communicate ideas for possible devices or solutions, 4.) build and test devices or solutions, 5.) determine if the devices or solutions solved the problem and refine the design if needed, and 6.) communicate the results.

LESSON OBJECTIVES:

TSW weather instruments psychrometers, hygrometers, and barometers to a) measure weather conditions, b) make weather predictions, and c) create weather forecasts. TSW use algorithms to calculate various equations to find relative humidity. TSW use calculations to locate information on data tables and grids. TSW use data to construct explanations of weather events.

ENGAGEMENT: IN THE CLASSROOM

- Teacher will discuss the root word, *meter*, which means measure.

MATERIALS:

Teacher Preparation

- Divide students into four groups and in pairs within each group. Due to the abundance of materials, the small group arrangements, and the sequence of detailed procedures, adult assistants or volunteers would be helpful to complete this lesson in a timely manner. One adult should lead each group, so the adults need to be familiar with the procedures for constructing the weather instruments ahead of time. If no volunteers are available, then it would be best to spread these lessons over several days. For the Hair Hygrometer Group, cut out the triangular pointers ahead of time. Use thin plastic and cut a right triangle about six inches long.

For each student

- Science notebooks
- Clipboard
- For the class:
- Chart paper
- Markers
- Sticky notes
- Index cards
- Tape
- Glue
- Scissors

SLING PSYCHROMETER GROUP
(Advanced Learners)

- Two student thermometers
- String or twine about 3 feet in length
- Cotton shoelaces ½ in. pieces
- Cup of water
- Safety goggles
- Calculator
- Two rubber bands
- Stopwatch or timer

HAIR HYGROMETER GROUP
(Advanced and On-Level Learners)

- Piece of heavy cardboard manila file folder
- 1 strand of human hair at least 9 inches long
- 2 pushpins
- 1 piece of thin plastic
- Scissors
- Tape
- Ruler

BAROMETER GROUP 1
(On-Level Learners)

- Clean, empty soup can with top removed
- Beverage can
- Balloon or plastic wrap
- Rubber band
- Plastic straw 8 in.
- Glue
- Index cards
- Masking tape

BAROMETER GROUP 2
(Below-Level Learners)

- Two-liter plastic bottle clear
- Food coloring
- Clear plastic cup
- liter of water
- Permanent marker
- Cm ruler
- *Hygrometers, Psychrometers and Barometers* student investigation guide (attached)
- *Hygrometers, Psychrometers and Barometers* teacher guide (attached)

EXPLORATION:

- Teacher will explain to student what psychrometers, hygrometers, and barometers measure and how they work.
- Teacher will also explain that students will be working with a group to design and construct one of these devices they will then use the device to collect weather data in order to make predictions about weather conditions.

EXPLANATION:

- Using the *Hygrometers, Psychrometers and Barometers* teacher guide, the teacher will guide students through the completion of the *Hygrometers, Psychrometers and Barometers* student investigation guide.

ELABORATION: IN THE CLASSROOM

- Students will continue to monitor and collect weather data from the garden.

EVALUATION: IN THE CLASSROOM

- Students will turn in the *Hygrometers, Psychrometers and Barometers* student investigation guide.
- Teacher will observe reactions and interactions with other students and the measurement tools.

RESOURCES:

1. *Hygrometers, Psychrometers and Barometers* handout (attached).

TEACHER GUIDE

HYGROMETERS, PSYCHROMETERS AND BAROMETERS

BACKGROUND:

Students should understand that measuring weather conditions is an important skill for predicting and monitoring conditions in the garden.

ENGAGEMENT:

Teacher will introduce new vocabulary psychrometer, hygrometer, and barometer.

- Ask, "**What do these words have in common?**" They share a common root word: *meter*. **What does meter mean as a root word measure ?** List other words on the board that contain the same root meter , such as speedometer, odometer, thermometer, and pedometer.
- Explain: Today, you are going to learn what psychrometers, hygrometers, and barometers measure and how they work. You will also work with a group to design and construct one of these devices. Then you will use the device to collect weather data in order to make predictions about weather conditions.

EXPLORATION:

- Focus Questions:
  –**How can we use weather instruments in the garden to predict precipitation?**
  –**What are psychrometers, hygrometers, and barometers, and how do they work?**
- A *hygrometer* is a weather instrument that measures moisture content in the air humidity. "Hygro" is a root word that means wet or moist. The warmer an air mass is, the more moisture it can hold; and the colder an air mass is, the less moisture it can hold. Have you ever had a bad hair day when the air is humid? When humidity increases, the length of hair strands also increase. Curly hair tends to get frizzy and straight hair becomes limp. In 1783, meteorologists used this observation to create a hair hygrometer to measure changes in humidity.
- A *psychrometer* is a weather instrument that measures relative humidity, dew point, and air temperature. "Psychro" is a root word that means cold. A sling psychrometer is a hygrometer with two thermometers, a wet-bulb thermometer and a dry-bulb thermometer. When using measurements in conjunction with the accompanying charts, they provide the relative humidity, the dew point, and the temperature. To facilitate cooling of the wet bulb, the two thermometers are vigorously spun through the air. Use caution when spinning the sling psychrometer in

order to avoid injury. Spin the psychrometer for about 30 seconds. The difference between the two thermometers helps determine the relative humidity and the dew point of the air. Further instructions about how to use the sling psychrometer are detailed in the lesson plan.

- A *barometer* is a weather instrument that measures the amount of pressure the atmosphere exerts on the earth. "Baro" is a root word that means pressure. This air pressure changes constantly and the direction in which it changes helps weather forecasters predict weather conditions. Barometers indicate the current barometric pressure as well as how the pressure changed since the last reading. Increasing air pressure suggests that fair weather is on the way, and decreasing air pressure indicates that clouds and / or precipitation are on the way.

EXPLANATION:

- Divide the class into four groups: Sling Pyschrometer Group (above-level), Hair Hygrometer Group (above and on-level), Barometer Group 1 (on-level), and Barometer Group 2 (on and below-level).

SLING PSYCHROMETER GROUP:

- Place the two thermometers back to back and securely fasten the thermometers together with a rubber band. Thread some string through the thermometers to create a necklace, so you can hang the sling psychrometer when not in use. Wrap the bulb of one thermometer with a piece of shoelace sock and secure the sock to the bulb with a rubber band. The thermometer without the sock is the dry-bulb thermometer and the one with the sock is the wet-bulb thermometer.
- Saturate the sock of the wet-bulb thermometer with water. Spin the psychrometer overhead for about 30 seconds. After spinning the thermometers, record the readings of each one. Find the wet-bulb depression by calculating the difference between the dry/wet-bulb thermometers. Repeat the highlighted steps ten times and record the wet-bulb depression after each trial.
- Alternate students with each trial so everyone gets to participate.
- Calculate the average mean of the ten wet-bulb depression readings. Then use the wet-bulb depression mean and the accompanying tables to determine the relative humidity and dew point of the air. Students repeat this sequence for several days and monitor the data in their science notebooks.

HAIR HYGROMETER GROUP:

- Cut a right triangle pointer about six inches in length from a thin piece of plastic. Situate the pointer so the right angle faces up. Cut two slits at the bottom of the pointer about one inch from the left edge. Tape a dime onto the pointer about two inches from the left edge. Cut two slits on the cardboard's top edge about an inch apart and one inch from the left side. Attach the pointer to the cardboard with a push pin about an inches from the left edge and about three quarters of the way down the side. Attach the hair strand by inserting it through the slit at the top of the cardboard and at the bottom of the pointer. Tape the hair securely into both sets of slits so that it is perpendicular to the top of the pointer. Push the pin through the pointer so that the top of the pointer is horizontal when the hair is slightly stretched. Gently wiggle the pointer up and down to make sure it can move.

- Take the completed hair hygrometer inside the school building to observe changes in the pointer.
- Encourage students to research the relative humidity on the National Weather Service website for one week.
- Each time, they draw a line at the tip of the arrow on their hair tool and record the actual humidity. These marks provide relative units for the hair hygrometer.

BAROMETER GROUP 1:

- Stretch a layer of balloon material or plastic wrap over the lid of the jar or coffee can and seal it with a rubber band so the container is airtight. Place a plastic straw horizontally with two-thirds of it on the container. Attach the straw to the layer of plastic wrap with tape. Tape another straw to the end of the first one.
- Challenge students to figure out how to monitor and record the movement of the straw, using an index card and a pencil. They should observe how the straw rises and falls for several days.

BAROMETER GROUP 2:

- Pour two-three inches of water into the clear plastic cup and stir several drops of food coloring into the water. Place the two-liter bottle upside down into the cup so that the opening of the bottle is below the waterline but above the bottom of the cup. The bottle does not need to be airtight but it should fit securely into the cup. Insert a plastic straw between the outside of the bottle and the brim of the cup so air can enter the cup. Keep the bottom of the straw above the waterline. With a permanent marker, indicate the waterline on the side of the cup and on the side of the bottle.
- Students monitor the level of the water inside the cup and bottle for several days. Encourage students to measure the changes in centimeters above and below the original waterline and record the data in their science notebooks.

ALL GROUPS:

- Once student pairs complete their weather instrument construction, vote on which device we should attach to the Weather Station. Use wind zip ties to securely attach weather instruments to the Weather Station. Other student pairs select a location in which to attach their weather instruments.
- Ask students in each group to explain what s happening to their respective devices. **What changes do you notice?** Students will sketch diagrams of their devices in their science notebooks, including labels. Each group elects a student representative to describe what they made and explain how it works to the class.

SLING PSYCHROMETER AND HAIR HYGROMETER GROUPS:

- Explain to students that water exists in the air and atmosphere even though it is not always visible. Guide students to recall that water in a gaseous state is called water vapor. The amount of water vapor in the air at a given temperature is called relative humidity. The temperature at which water vapor condenses into liquid water is called dew point. The closer the range of air temperature to the dew point, the greater the percentage of precipitation. By measuring relative humidity and dew point, meteorologists make weather predictions and develop weather forecasts. Pose the following questions for students to discuss and respond to in their science notebooks. **How does the air feel when the humidity changes? How do plants respond to changes in humidity? What do changes in humidity indicate about weather forecasts?**

BAROMETER GROUPS 1 AND 2:

- Challenge students to develop a plan to measure air pressure changes over time and observe the corresponding weather patterns. Within the group, compare the barometers made by each pair. Ask: **How do you know if your barometer is accurate? Does it predict precipitation or storms? What weather patterns relate to the data collected from your barometer?**

ALL GROUPS:

- Return to the classroom as a whole group. Students create a table in their science notebooks to explain what each instrument does, how it works, and its importance to the garden. They should sketch an example of each student-constructed device as well as an example of a standard manufactured device, and write conjectures for using barometer #2.
- Fair weather results in high air pressure and pushes the water level down in the cup and up in the bottle. Inversely, stormy weather results in low air pressure and pushes the water level up in the cup and down in the bottle.
- Students will complete the *Hygrometers, Psychrometers and Barometers* student investigation guide.
- Vocabulary: *relative humidity, thermometer, psychrometer, hygrometer, barometer, air pressure, temperature, moisture, water vapor, atmosphere, troposphere, meteorology, forecasting, forecast, weather instruments, design, engineer, modify, optimize, dew point, wet-bulb depression, difference, sum, average, mean.*

ELABORATION:

- Ask: **How can we use the data we collect to make more accurate weather forecasts?**
- Encourage students to develop a plan to document and display the data gathered over time.
- Challenge students to relate weather data to observable weather patterns and to use data to inform decisions about garden preparation and maintenance. Each quarter, switch groups to allow other students to explore the use of other weather instruments.

- Always encourage students to make predictions for each month. Likewise, have them record their "wonderings" on sticky notes for display to guide future investigations.

EVALUATION:

- Students will turn in the *Hygrometers, Psychrometers and Barometers* student investigation guide.
- Teacher will observe reactions and interactions with other students and the measurement tools.

HYGROMETERS, PSYCHROMETERS AND BAROMETERS

Name(s) ______________________________

Sketch your weather instrument design below.

List the materials you will need for construction.

____

____

____

FOURTH GRADE 5E LESSON PLAN

Be a Meteorologist

SUBJECT / GRADE LEVEL: Fourth Grade - Earth Science - Weather Station Creation Part 5.

ACADEMIC STANDARDS AND PERFORMANCE INDICATORS:

Science:

4.E.2B.1 Analyze and interpret data from observations, measurements, and weather maps to describe patterns in local weather conditions including temperature, precipitation, wind speed / direction, relative humidity, and cloud types and predict changes in weather over time.

4.E.2B.3 Construct explanations about regional climate differences using data from the long term weather conditions of the region.

4.S.1A.4 Analyze and interpret data from informational texts, observations, measurements, or investigations using a range of methods such as tabulation or graphing to 1.) reveal patterns and construct meaning or 2.) support explanations, claims, or designs.

4.S.1A.7 Construct scientific arguments to support claims, explanations, and designs using evidence from observations, data, or informational texts.

LESSON OBJECTIVES:

TSW use the data they collected for one week to write a Weekend Weather Bulletin to forecast the weather conditions for the upcoming weekend. TSW include a weather map of South Carolina and use symbols to show frontal movements, high and low pressure areas, severe weather warnings, temperatures, chance of precipitation, and wind speed and direction. They will use a rubric as a guide for their writing and will present their Weather Bulletin orally to their classmates.

ENGAGEMENT:

- Teacher will post the question, "**What does a meteorologist do?**"

EXPLORATION:

- Teacher will show students a video clip of the weather forecast from a news broadcast.

MATERIALS:

- *Weather Data Table* (attached)
- Student Weather Map map of the Southeast
- Blank map of the area where you are located/city www.maps.google.com
- Newspapers
- *Weather Forecast Symbols* http://www.opc.ncep.noaa.gov/product_description/keyterm.shtml
- Hi-lighters two colors
- Markers or colored pencils blue, red, yellow, and black
- Poster board
- Glue
- Tape
- Scissors
- *Be A Meteorologist* student investigation guide (attached)
- *Be A Meteorologist* teacher guide (attached)

EXPLANATION:

- Teacher will introduce weather map symbols, define their meanings, and relate each symbol to the instruments from the Weather Station.

ELABORATION:

- Using the *Be A Meteorologist* teacher guide, the teacher will guide students through the completion of the *Be A Meteorologist* student investigation guide.
- Students will continue to monitor instruments and collect weather data from the garden.

EVALUATION:

- Students will turn in the *Be A Meteorologist* student investigation guide.
- Teacher will observe reactions and interactions with other students and the weather maps.

RESOURCES:

1. *Weather Data Table* (attached).
2. *Weather Measurement Tools:* student example (attached).
3. *Weather Forecast Symbols* http://www.opc.ncep.noaa.gov/product_description/keyterm.shtml.
4. *Be a Meteorologist* handout (attached).

TEACHER GUIDE
BE A METEOROLOGIST

BACKGROUND:

Students should understand that meteorologists gather data collected from weather instruments and report their findings to the public.

TEACHER PREPARATION:

This 2 day lesson is the summative assessment for the Weather Station Creation Unit. This lesson uses the data collected over the course of one week from scientific instruments in the garden, but the lesson does not need to take place in the garden. It can, however, be used in conjunction with some garden based data collection or farm maintenance. Students make a brochure Weekend Weather Bulletin to forecast the weather conditions for the upcoming weekend and present their predictions to their classmates. It is more meaningful if students were able to present to families or any volunteers who helped them during the unit. It is also fun to conduct the oral presentations in the garden in view of the Weather Station.

ENGAGEMENT:

- Teacher will post the question, "**What does a meteorologist do?**"

EXPLORATION:

- Ask: **Who has ever seen a weather forecast on the local news?**
- Encourage students to explain what purpose a weather forecast from a meteorologist serves.
- Show students a video clip from a local meteorologist.
- Ask them to identify the key features to the broadcast and list them on chart paper high/low temperatures, wind speed/direction, humidity, wind chill, chance of precipitation, clouds, outdoor activities, weather maps, etc...
- Explain the lesson objectives and outcomes to students. Focus Questions:
  –**How can we use weather data to create a weather map and write a weather forecast?**
  –**How is weather data useful to gardeners and farmers?**
- Group students with a partner. Allow 10–15 minutes for students to use weather measurement instruments to record data for the following: air temperature, relative humidity, wind speed, wind direction, precipitation, cloud types, cloud movements, cloud coverage, sunlight/visibility, and air pressure. They record data on a *Weather Data Table* (attached). Insist that both partners

read each device to make sure they record accurate readings. Once all students collect data, return to the classroom.

- Students work in pairs to analyze data. They color-code qualitative data and quantitative data with two different colors of hi-lighter pens.
- Give each pair of students the weather section of a local newspaper and encourage them to read it to: a. Compare data from the newspaper to data we recorded from the garden. b. To make predictions about upcoming weather conditions.

EXPLANATION:

- Introduce weather map symbols, define their meanings, and relate each symbol to the instruments from the Weather Station. Play a video clip of a weather forecast, pausing several times to identify symbols, such as fronts, air pressure, temperature, and humidity. Distribute Student Weather Maps map of the Southeast. Draw a compass rose on the weather map and review directions with students.
- Give the following oral directions for student weather map construction:

  1. Draw black marker a low pressure symbol centered just north of Charlotte, North Carolina.
  2. Draw blue marker a cold front that begins near Lexington, Kentucky and extends toward the Northeast.
  3. Draw red marker a warm front that begins near Atlanta, Georgia and extends toward the Southeast through Savannah, Georgia and out into the Atlantic Ocean.
  4. Use a red marker to draw a severe thunderstorm warning box around Columbia, South Carolina.
  5. Use a yellow marker to draw a severe thunderstorm watch box around Asheville, North Carolina.
  6. Use a black marker to draw a wind feather for Charleston, West Virginia that shows clear skies with wind coming from the south at 15 MPH.
  7. Use a black marker to draw a wind feather for Roanoke, Virginia that shows partly cloudy skies with wind coming from the southwest at 20 MPH.
  8. Use a black marker to draw a wind feather Knoxville, Tennessee that shows cloudy skies with wind coming from the west at 10 MPH.
  9. Ask the following questions:

  –Which city is under a severe thunderstorm warning and why?
  –Which city is under a severe thunderstorm watch and why?
  –What direction is the cold front moving?
  –How does the symbol show direction of the frontal movement?

BREAK: Continue the lesson tomorrow.

- Review the weather map symbols from yesterday s lesson. If necessary, meet with students who need help with the weather maps they created yesterday.
- Vocabulary: *meteorology*, *meteorologist*, *prediction*, *weather map*, *weather forecast*, and all unit vocabulary pertaining to weather conditions and scientific instruments, *qualitative*, *quantitative*.

ELABORATION:

- Students will complete the *Be A Meteorologist* student investigation guide.
- Today, students begin independently creating their *Weekend Weather Bulletins* on poster board. *The Weather Bulletin* forecast must include:
  –A weather map of our area, including fronts and air pressure symbols.
  –A chance of precipitation and / or severe weather.
  – Temperature and humidity.
  –Wind speed and direction.
  –A "what to wear" suggestion.
  –A statement about clouds.
  –A week s worth of weather data, including weekly averages. They may include a "message to farmers and gardeners".
- Circulate the classroom to assist students as needed.

EVALUATION:

- Students will turn in the *Be A Meteorologist* student investigation guide.
- Teacher will observe reactions and interactions with other students and the weather maps.

5.

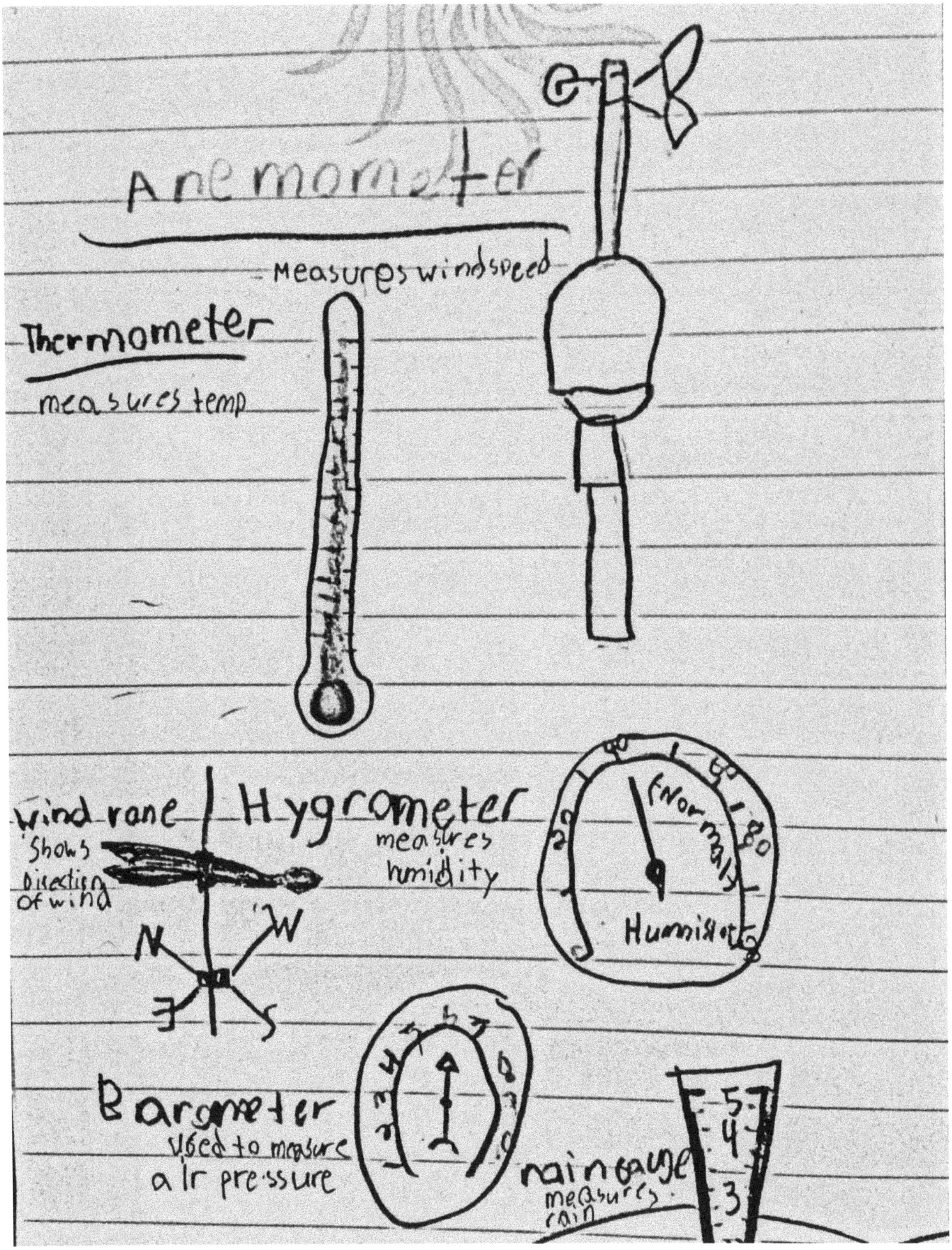
Anemometer
Measures windspeed
Thermometer
measures temp
wind vane
Shows Direction of wind
N
W
E
S
Hygrometer
measures humidity
Normal
Humidity
Barometer
used to measure air pressure
rain gauge
measures rain
5
4
3

Name ____________________ Partner Name________________ Class of ____________

WEATHER DATA TABLE FOR WRITING A WEATHER BULLETIN

Date/ Time	
Air Temperature	
Wind Direction	
Wind Speed	
Relative Humidity	
Precipitation	
Air Pressure	
Cloud Types	
Cloud Movements	
Cloud Coverage	
Sunlight/ Visibility	

BE A METEOROLOGIST

Name

Use this guide to collect the information for your Weather Bulletin forecast.

What is the chance of precipitation and/or severe weather?

What is the temperature and humidity?

What is the wind speed and direction?

What is your "what to wear" suggestion?

What is the wind speed and direction?

What is the cloud cover?

List remaining weather data, including weekly averages.

FOURTH GRADE LESSON PLAN

Head in the Clouds

SUBJECT / GRADE LEVEL: Fourth Grade - Earth Science.

ACADEMIC STANDARDS AND PERFORMANCE INDICATORS:

Science:

4.E.2B.1 Analyze and interpret data from observations, measurements, and weather maps to describe patterns in local weather conditions including temperature, precipitation, wind speed / direction, relative humidity, and cloud types and predict changes in weather over time.

4.S.1A.4 Analyze and interpret data from informational texts, observations, measurements, or investigations using a range of methods such as tabulation or graphing to 1.) reveal patterns and construct meaning or 2.) support explanations, claims, or designs.

LESSON OBJECTIVES:

TSW identify different types of clouds and classify them as cirrus, stratus, and cumulus. TSW use observations to make predictions about weather events. TSW recognize and explain characteristics of different cloud types and describe how clouds form.

MATERIALS:

- *The Cloud Book* by Tomie DePaola one copy for every two students
- *Head In The Clouds* student investigation guide (attached)
- *Head In The Clouds* teacher guide (attached)

ENGAGEMENT:

- Teacher will post the question, "**How were ancient people able to predict weather conditions before there were modern scientific instruments and tools?**"

EXPLORATION:

- Teacher will read aloud pages 20-27 of *The Cloud Book* by Tomie DePaola and discuss the history of weather section.

EXPLANATION:

- Using the *Head In The Clouds* teacher guide, the teacher will guide students through the completion of the *Head In The Clouds* student investigation guide.

ELABORATION:

- Students will continue to monitor clouds and collect data for several days.

EVALUATION:

- Students will turn in the *Head In The Clouds* student investigation guide.
- Teacher will observe reactions and interactions with other students and the clouds.

RESOURCES:

1. *The Cloud Book* by Tomie DePaola.
2. *Head in the Clouds* handout (attached).

TEACHER GUIDE
HEAD IN THE CLOUDS

BACKGROUND:

Students should understand that weather patterns can be predicted through close observation of the clouds.

ENGAGEMENT:

- Teacher will post the question, "**How were ancient people able to predict weather conditions before there were modern scientific instruments and tools?**"

EXPLORATION:

How can we use our senses to forecast the weather?

- Read aloud pages 20-27 of *The Cloud Book* by Tomie DePaola. Discuss the history of weather section.
- Focus Questions:
  –**What are clouds?**
  –**What do cloud shapes indicate about weather patterns?**
  –**How can we use cloud observations to make predictions about weather?**
  –**Why are weather predictions important to farming and agriculture?**
- Divide students into groups of three and have them find an observation spot where they have a view of the sky. In their science notebooks, students a. record observations about the weather conditions using qualitative data, including heat, wind, clouds, etc. , b. draw sketches of the different types of clouds they observe, and c. describe the characteristics of each type of cloud using adjectives and figurative language.
- Combine into groups of six and have them share science notebook entries with each other. As students communicate their entries, the teacher confers with each group to ask questions and make inferences. Explain to students that they will make cloud observations every day of the school year in order to accurately predict weather patterns and interpret changing weather conditions.

EXPLANATION:

- *Cumulus*: Puffy, lumpy-looking clouds often with a flat bottom. When cumulus clouds are dark they usually bring rain; white cumulus clouds do not bring rain.

- *Cirrus*: High, thin, wispy clouds. They are formed mostly of ice crystals. Cirrus clouds are most often associated with fair weather.
- *Stratus*: Layers of clouds that spread out covering a large area. Stratus clouds are often lower in the sky.
  ****The formation of clouds happens when water vapor in the air rises, cools and condenses or moves from a warm place to a cool place and condenses, forming the water droplets that make up a cloud. A cloud is a collection of tiny, liquid water droplets not water vapor gas.*
- Reinforce the concept that prior to modern data collection equipment; people used observations to predict the weather. Discuss problems with relying on the senses to forecast the weather i.e. hurricanes as well as problems with relying on technological devices (i.e. network outages). Ask: **Where do clouds form?** Discuss and define atmosphere and troposphere. Explain that almost all weather and cloud formation occurs in a layer of gases surrounding the surface of Earth called the *troposphere*.
- Cloud Graphics: Small groups create a graphic representation of clouds using the following requirements:
  –A visual representation of the three major cloud types.
  –A written description of their physical appearance and composition.
  –A statement on its relative location in the sky.
  –A list of adjectives and / or figurative language examples to describe each type of cloud.
- The graphic may contain any of the preceding elements for additional types of clouds and an explanation about what types of weather follow each type of cloud.
- Students observe clouds over the next several days until they witness examples of each cloud variety. Each time they see a new type of cloud, they complete their *Heads in the Clouds* handout (attached) , including the date / time, a statement about current weather conditions, and a prediction about weather conditions to come.
- Vocabulary: *clouds*, *cirrus*, *cumulus*, *stratus*, *technology*, *troposphere*, *atmosphere*, *altitude*, *horizon*, *direction*.

ELABORATION:

- Students will complete the *Head In The Clouds* student investigation guide over a period of several days.
- As students commit to making daily cloud observations, encourage them to make connections about how different types of clouds help us predict different types of weather.

EVALUATION:

- Students will turn in the *Head In The Clouds* student investigation guide.
- Teacher will observe reactions and interactions with other students and the clouds.

HEAD IN THE CLOUDS

Name

____

Complete the following chart based on your cloud observations.

Cirrus		
Date/Time	*Current Weather Conditions*	*Weather Prediction*
Sketch		

FOURTH GRADE LESSON PLAN
Seed Dispersal

SUBJECT / GRADE LEVEL: Fourth Grade - Life Science.

ACADEMIC STANDARDS AND PERFORMANCE INDICATORS:

Science:

4.L.5B.2 Construct explanations for how structural adaptations such as roots, stems, or leaves; color of flowers, or seed dispersal allow plants to survive and reproduce.

LESSON OBJECTIVES:

TSW about methods of seed dispersal and how the physical characteristics of seeds often determine how they are dispersed. TSW locate seeds from the garden and use the seed structure to infer how they are dispersed.

MATERIALS:
- *A Seed is Sleepy* by Dianna Hutts Aston
- *Seed Dispersal* student investigation guide (attached)
- *Seed Dispersal* teacher guide (attached)

ENGAGEMENT:
- Teacher will post the question, "**What seeds do we eat?**"
- Teacher will follow the engagement question by reading *A Seed is Sleepy* by Dianna Hutts Aston to the class.

EXPLORATION:
- Teacher will introduce and describe different methods of seed dispersal.

EXPLANATION:
- Using the *Seed Dispersal* teacher guide, the teacher will guide students through the completion of the *Seed Dispersal* student investigation guide.

ELABORATION:

- Students will search for different types of seeds and discuss their methods of dispersal.

EVALUATION:

- Students will turn in the *Seed Dispersal* student investigation guide.
- Teacher will observe reactions and interactions with other students and the seeds.

RESOURCES:

1. *A Seed is Sleepy* by Dianna Hutts Aston.
2. *Seed Dispersal* handout (attached).

TEACHER GUIDE
SEED DISPERSAL

BACKGROUND:

Students should understand that plants use a variety of methods for successful seed dispersal.

ENGAGEMENT:

- Teacher will post the question, "**What seeds do we eat?**"
- Teacher will follow the engagement question by reading *A Seed is Sleepy* by Dianna Hutts Aston to the class.

EXPLORATION:

- Explain that humans are frequently the way seeds get from one place to another. However, humans are not the only way seeds can move from place to place. Ask students **why a seed would need to move away from its parent plant. What are some ways that they do this?**
- Most plants produce a large number of seeds because some seeds do not survive.
- In order to ensure that seeds will survive; they must be carried away dispersed from the parent plant.
  –Some seeds have hooks on them that allow them to attach to animal fur or clothes.
  –Some seeds are able to float in water.
  –Some seeds are light and have wings or thin hairs that allow them to be carried away by wind.
  –Some seeds are eaten by animals and deposited in areas away from the parent plants.
- Seeds stand a better chance of growing if they do not have to compete with the parent plant for light, water and nutrients. Since plants do not move, fruits and seeds have developed unique structures to help them move to a new place where there is less competition and the growing conditions are more favorable for growth and survival.

EXPLANATION:

- Ask students to recall ways that seeds are dispersed from the book and make a list floating in water, hitchhiking on another organism, carried by the wind, being eaten then deposited.
- Instruct students to search for 3 more seeds and encourage them to search for seeds that are not adapted to attach to an animal s fur or person s clothing.
- If students suggest that the shape of a seed would cause it to be dispersed by the wind, encourage them to blow on them or toss them in the air then observe.

- If students suggest that the shape of a seed would cause it to float in water, allow them to test the seed's buoyancy in a small bucket of water.
- If students find a tasty fruit that may be eaten by an animal, cut the fruit open and allow students to observe the seeds inside.
- Bring students back together. Once students have time to share their findings, pose the following questions:
  –What characteristics do seeds have that allow them to be moved to another location to grow?
  –Why is seed dispersal important to a plant?
  –Why couldn't all seeds just drop by the parent plant and grow there? What would be the disadvantage?
- Vocabulary: *parent plant*, *seed*, *dispersal*, *reproduction*, *pollination*, *physical characteristics*, *survival*, *buoyancy*, *floating*, *sinking*.

ELABORATION:

- Students will complete the *Seed Dispersal* student investigation guide.
- Seed Challenge:
  1. Divide the class into small groups. Review seed dispersal strategies. Challenge each team to invent a strategy for seed dispersal. The seed might float on wind, explode, hitchhike, float, be eaten by a bird or animal, or some other mechanism that they develop.
  2. Supply students with a variety of miscellaneous materials to help them create the new seed dispersal. These materials should be used to modify seeds. Provide a number of different seeds for them to use. Let them know that they will have a half an hour to build their seed dispersal mechanism. Suggest that each team draw their separate design and then the group should decide on one strategy to build 25 minutes.
  3. Ask teams to develop and write a description for how their seed will travel 10 minutes.
- Testing:
  1. Ask each group to demonstrate their seed dispersal mechanisms by launching it towards a six inch pot. Give each team three try to reach the pot. Offer teams a chance to modify their design to improve success.

EVALUATION:

- Students will turn in the *Seed Dispersal* student investigation guide.
- Teacher will observe reactions and interactions with other students and the seeds.

Seed Collection Chart

Seed #	Characteristics (Description of Seed)	Sticks to Things?	Moved by Wind?	Floats in Water?	Possible Methods of Dispersal?	Characteristic I used to determine how it may move?
1						
2						
3						
4						

SEED DISPERSAL

Name ______________________________

Complete the following chart based on methods of seed dispersal.

Characteristics	*Method of Dispersal*
	Wind
	Water
	Animals
	Humans

How does seed dispersal affect plant survival?

____

____

____

____

FOURTH GRADE LESSON PLAN
Structural Adaptations

SUBJECT / GRADE LEVEL: Fourth Grade - Life Science.

ACADEMIC STANDARDS AND PERFORMANCE INDICATORS:

Science:

4.L.5B.3 Construct explanations for how structural adaptations such as methods for defense, locomotion, obtaining resources, or camouflage allow animals to survive in the environment.

LESSON OBJECTIVES:

TSW analyze qualitative observations to draw conclusions for how insects use their bodies to survive in their habitat.

MATERIALS:

- Hand lenses
- iPad (optional)
- *Structural Adaptations* student investigation guide (attached)
- *Structural Adaptations* teacher guide (attached)

ENGAGEMENT:

- Teacher will post the question, "**What structures do humans have that help them to pick up and eat food?**"
- Teacher will follow up the engagement question with this video: *Adaptations* by Mark Drollinger https://www.youtube.com/watch?v=-IiP-9VzY9w.

EXPLORATION:

- Teacher will instruct students to observe pollinators in the garden noticing their structural adaptations.

EXPLANATION:

- Using the *Structural Adaptations* teacher guide, the teacher will guide students through the completion of the *Structural Adaptations* student investigation guide.

ELABORATION

- Students can create an electronic or paper book of adaptations highlighting organisms found in the garden.

EVALUATION:

- Students will turn in the *Structural Adaptations* student investigation guide.

RESOURCES:

1. *Adaptations* by Mark Drollinger https://www.youtube.com/watch?v=-IiP-9VzY9w.
2. *Structural Adaptations* handout (attached).
3. *Structural Adaptations* student example (attached).

TEACHER GUIDE

STRUCTURAL ADAPTATIONS

BACKGROUND:

Students should understand that adaptations are characteristics that help living things survive in their habitat.

ENGAGEMENT:

- Teacher will post the question, "**What structures do humans have that help them to pick up and eat food?**"
- Teacher will follow up the engagement question with this video: *Adaptations* by Mark Drollinger: https://www.youtube.com/watch?v=-IiP-9VzY9w.

EXPLORATION:

What structural adaptations do the insects in our garden use and how do these structures help them locate flowers, gather nectar, defend themselves, and collect then distribute pollen?

- Encourage students to use their hand lenses to closely observe how the insects are moving from flower to flower, landing, obtaining nectar, defending themselves, and collecting pollen then distributing pollen to other flowers or other parts of the same flower.
- Instruct students to record a labeled illustration of their observation in their science journals. Students may also collect insects in small containers to observe, illustrate, and label in a controlled setting. Students may make detailed observations over several days, during various times of day and weather.

EXPLANATION:

- Students will complete the *Structural Adaptations* student investigation guide.
- Students will share their observations and confirm or disprove each other's theories through discussion. The teacher will facilitate the discussion asking students to provide evidence that support their claims.
- Vocabulary: *structures*, *adaptations*, *survival*, *obtain*, *locomotion*, *defense*.

ELABORATION:

- Students can use book creator on their iPads or another organization model to create a book on the topic. Their books may be broken into sections that highlight multiple types of pollinators such as wasps, bats, and hummingbirds. The sections may contain an illustration or image along with a detailed description of how the insect uses its structures to adapt to its habitat.

EVALUATION:

- Students will turn in the *Structural Adaptations* student investigation guide.

12

What structured adaptions do the Insects in our garden use and How do these structures Help them Survive?

April 8, 2014
10:27AM

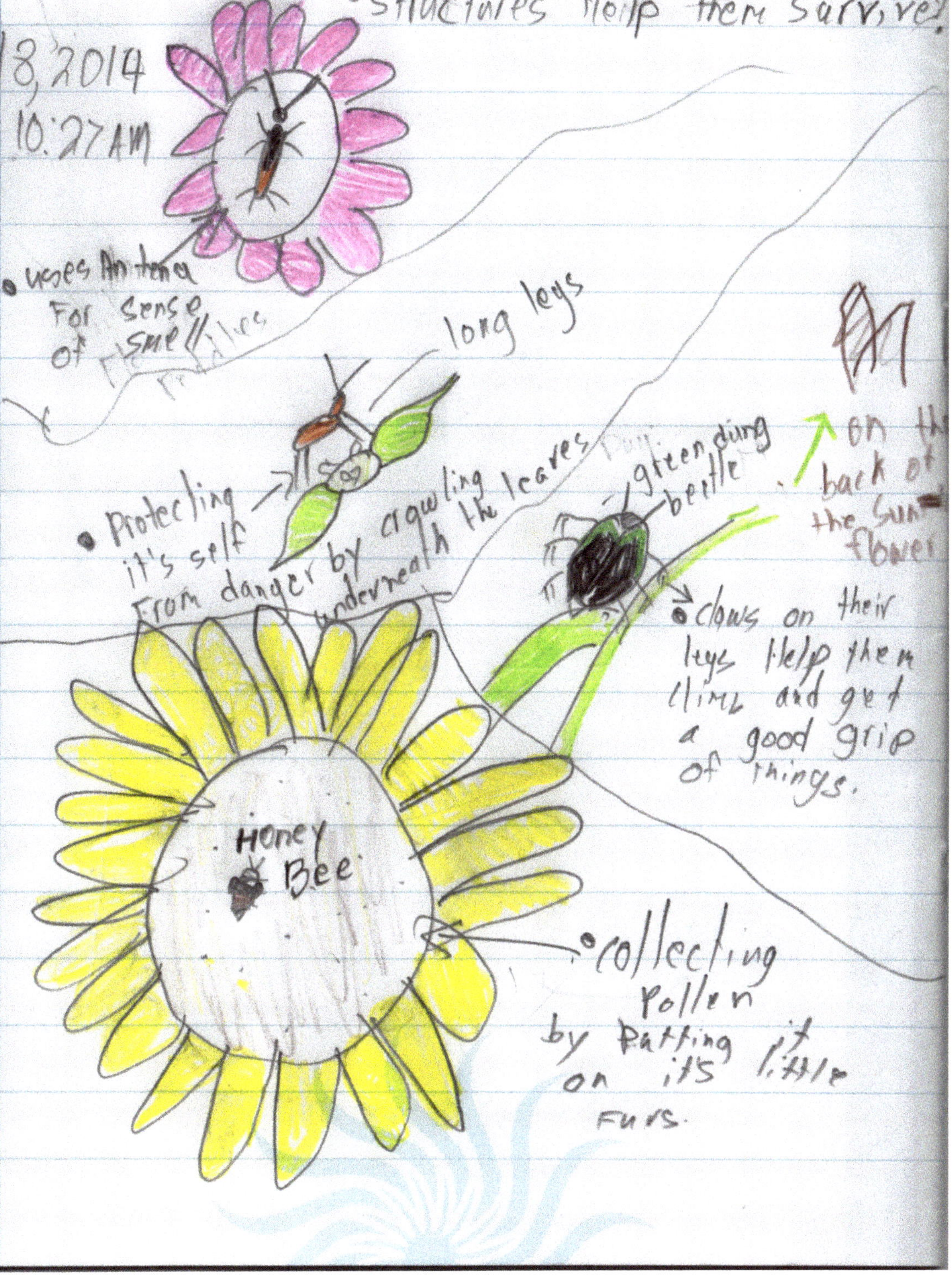

STRUCTURAL ADAPTATIONS

Name

___

Complete the following chart based on observed adaptations in the garden.

Structural Adaptation 1 Sketch	Structural Adaptation 2 Sketch
Structural Adaptation 1 Description	**Structural Adaptation 2 Description**

FOURTH GRADE LESSON PLAN
Final Scavenger Hunt

SUBJECT / GRADE LEVEL: Fourth Grade - Inquiry and Life Science.

ACADEMIC STANDARDS AND PERFORMANCE INDICATORS:

Science:

4.L.5B.1 Develop and use models to compare how humans and other animals use their senses and sensory organs to detect and respond to signals from the environment.

4.L.5B.2 Construct explanations for how structural adaptations such as roots, stems, or leaves; color of flowers, or seed dispersal allow plants to survive and reproduce.

4.L.5B.3 Construct explanations for how structural adaptations such as methods for defense, locomotion, obtaining resources, or camouflage allow animals to survive in the environment.

4.E.2B.1 Analyze and interpret data from observations, measurements, and weather maps to describe patterns in local weather conditions including temperature, precipitation, wind speed / direction, relative humidity, and cloud types and predict changes in weather over time.

4.S.1A.4 Analyze and interpret data from informational texts, observations, measurements, or investigations using a range of methods such as tabulation or graphing to 1.) reveal patterns and construct meaning or 2.) support explanations, claims, or designs.

4.E.2B.3 Construct explanations about regional climate differences using data from the long term weather conditions of the region.

4.S.1A.7 Construct scientific arguments to support claims, explanations, and designs using evidence from observations, data, or informational texts.

4.S.1A.3 Plan and conduct scientific investigations to answer questions, test predictions and develop explanations: 1.) formulate scientific questions and predict possible outcomes, 2.) identify materials, procedures, and variables, 3.) select and use appropriate tools or instruments to collect qualitative and quantitative data, and 4.) record and represent data in an appropriate form. Use appropriate safety procedures.

4.S.1B.1 Construct devices or design solutions to solve specific problems or needs: 1.) ask questions to identify problems or needs, 2.) ask questions about the criteria and constraints of the devices or solutions, 3.) generate and communicate ideas for possible devices or solutions, 4.) build and test devices or solutions, 5.) determine if the devices or solutions solved the problem and refine the design if needed, and 6.) communicate the results.

LESSON OBJECTIVES:

TSW include a weather map of South Carolina and use symbols to show frontal movements, high and low pressure areas, severe weather warnings, temperatures, chance of precipitation, and wind speed and direction. TSW construct weather instruments to measure wind direction and wind speed and determine the helpful and harmful ways wind affects the garden. TSW identify different types of clouds and classify them as cirrus, stratus, and cumulus. TSW use observations to make predictions about weather events. TSW recognize and explain characteristics of different cloud types and describe how clouds form. TSW weather instruments psychrometers, hygrometers, and barometers to a measure weather conditions, b make weather predictions, and c create weather forecasts. TSW use algorithms to calculate various equations to find relative humidity. TSW use calculations to locate information on data tables and grids. TSW use data to construct explanations of weather events. TSW about methods of seed dispersal and how the physical characteristics of seeds often determine how they are dispersed. TSW locate seeds from the garden and use the seed structure to infer how they are dispersed. TSW use a rain gauge constructed from reusable materials to measure amounts of precipitation and monitor responsible irrigation practices using data collected from the homemade rain gauges. TSW analyze qualitative observations to draw conclusions for how insects use their bodies to survive in their habitat. TSW measure soil temperatures to determine the best times to plant vegetable crops in the garden. TSW measure daily high and low air temperatures, find the range, and plot the data on a line graph. TSW determine the number of days in a given month there are for maximum plant growth, using thermo-period data. TSW compare air temperature in direct sunlight to air temperature in shade to make inferences about plant growth.

MATERIALS:

- *Catching Air* student investigation guide
- *Head In The Clouds* student investigation guide
- *Hygrometers, Psychrometers and Barometers* student investigation guide
- *Be A Meteorologist* student investigation guide
- *Seed Dispersal* student investigation guide
- *Precipitation or Irrigation? Let's Take a Rain Check* student investigation guide
- *Structural Adaptations* student investigation guide
- *The Garden Is So Hot Right Now* student investigation guide
- Paper lunch sacks, paper, and pencil to collect items

ENGAGEMENT:

- Teacher will pose the question, "**What is a term that you've learned during our garden lessons?**"

EXPLORATION:

- Teacher will poll students for plant terms until all of the terms for the scavenger hunt have been discussed.

EXPLANATION:

- Using the *Scavenger Hunt* chart, the teacher will guide students through the explanation of all terms in the chart.
- After reviewing the *Scavenger Hunt* chart, students will go to the garden in order to find the items.

ELABORATION:

- Students can compost all of the items from the scavenger hunt including the chart and the paper bag.
- If there is nowhere to compost the items, this could also be turned into a digital scavenger hunt where students simply take pictures of the items in the chart.

EVALUATION:

- Students will show teacher items in the bag as the teacher checks them off of the *Scavenger Hunt* chart.

RESOURCES:

1. *Scavenger Hunt* handout (attached).

SCAVENGER HUNT

Name

____

Instructions: Find all of the items listed below:

Item	Where did you find it?
1) Humidity reading from hygrometer	
2) Temperature reading from thermometer	
3) Wind speed reading from anemometer	
4) Sketch of Cirrus cloud	
5) Pressure reading from barometer	
6) A seed dispersed by wind	
7) A seed dispersed by water	
8) Precipitation reading from rain gauge	
9) Sketch of a structural adaptation found on a garden organism	
10) Garden soil temperature reading	

LESSONS FOR FIFTH GRADE

FIFTH GRADE LESSON PLAN
Maximizing Garden Yield

SUBJECT / GRADE LEVEL: Fifth Grade - Life Science.

ACADEMIC STANDARDS AND PERFORMANCE INDICATORS:

Science:

5.L.4A.1 Analyze and interpret data to summarize the abiotic factors including quantity of light and water, range of temperature, salinity, and soil composition of different terrestrial ecosystems and aquatic ecosystems.

5.L.4B.2 Obtain and communicate information to describe and compare the biotic factors including individual organisms, population, and communities of different terrestrial and aquatic ecosystems.

5.L.4B.4 Construct scientific arguments to explain how limiting factors including food, water, space, and shelter or a newly introduced organism can affect an ecosystem.

5.S.1A.3 Plan and conduct controlled scientific investigations to answer questions, test hypotheses and predictions, and develop explanations: 1.) formulate scientific questions and testable hypotheses, 2.) identify materials, procedures, and variables, 3.) select and use appropriate tools or instruments to collect qualitative and quantitative data, and 4.) record and represent data in an appropriate form. Use appropriate safety procedures.

LESSON OBJECTIVES:

TSW identify and describe the biotic, abiotic, and limiting factors of the garden ecosystem and classify and compare them according to their characteristics. TSW analyze and interpret data about the abiotic factors present in the garden to summarize the availability of resources necessary for optimum plant growth.

MATERIALS:

- Garden plot cleared and ready for planting
- Measuring tape
- Packets of one type of seeds to plant depending on planting season
- *Growth Data* and *Plant Type Data* tables (attached)
- *Maximizing Garden Yield* student investigation guide (attached)
- *Maximizing Garden Yield* teacher guide (attached)

ENGAGEMENT:

- Teacher will ask students to describe abiotic factors that affect plant growth.
- Teacher will prompt students as they respond until they identify water, light, temperature humidity, and nutrition.
- Review biotic, abiotic, and limiting factors with students, providing examples applicable to your garden.

EXPLORATION:

- Students will investigate how they can control the garden's abiotic factors to maximize the yield of what they are planting.

EXPLANATION:

- Students should predict how much produce they think they will be able to harvest.
- To do this, they need to calculate the average amount of produce that grows on one plant based on how many seeds were planted. Students should record this in their science notebooks under the focus question. You may provide a sentence starter:
  I predict our plants will yield __________________ insert type of produce and quantity here. I predict that the plants will yield this because...
- Finally, instruct students to record the planting suggestions and needs of the plant in their science notebooks including the estimated amount of time it will take for the plant to produce the fruit or vegetable.

ELABORATION: IN THE GARDEN

- Using the *Maximizing Garden Yield* teacher guide, the teacher will guide students through the completion of the *Maximizing Garden Yield* student investigation guide.

EVALUATION:

- Students will complete *Maximizing Garden Yield* student investigation guide.

RESOURCES

1. *Growth Data Table* (attached).
2. *Plant Types Data Table* (attached).
3. *Maximizing Garden Yield Cycles* handout (attached).

TEACHER GUIDE

MAXIMIZING GARDEN YIELD

BACKGROUND:

Students should understand that there are environmental factors which affect plant growth.

ENGAGEMENT:

- Teacher will ask students to describe abiotic factors that affect plant growth.
- Prompt students as they respond until they identify water, light, temperature humidity, and nutrition. Review *biotic*, *abiotic*, and *limiting factors* with students, providing examples applicable to your garden.

EXPLORATION:

- Explain that students will begin investigating three abiotic factors that affect plant growth: water, sunlight, and soil composition.
- Explain that during this lesson and throughout the year students will be working to generate solutions to feeding a rapidly increasing human population.
- **How can we control the garden's abiotic factors to maximize the yield of what we are planting?**
- Next, students should research the abiotic factors that affect what they will be planting in the garden.
- **How much sunlight does the plant need? How much water does it need daily?**
- **What soil composition is ideal for this type of plant? How much space does the plant need between other plants?** Students can do this online or by viewing planting instructions on the seed packets.
- **Ask students to recall how much space is needed between each seed planted and how deep the seed should be placed into the soil.**
- Next, instruct students to measure the length and width of the garden bed or beds to determine how many seeds they will plant students may work in pairs or groups to form a conclusion.
- Students will then follow the planting suggestions to place seeds in the soil. Make sure that students remember how many seeds were planted.

EXPLANATION:

- Using the *Maximizing Garden Yield* teacher guide, the teacher will guide students through the completion of the *Maximizing Garden Yield* student investigation guide.
- *Biotic Factors* are the living parts of an ecosystem, including populations and communities of organisms and microorganisms. *Populations* are all the members of one kind of organism that live in a particular area. *Communities* are a group of different populations of organisms. *Microorganisms* are single-cell or multi-cell organisms that are too small to be seen with the naked eye.
- *Abiotic Factors* are the nonliving parts of an ecosystem, including temperature range of temperature , water quantity, sunlight quantity, soil composition, and air gases.
- *Limiting Factors* are the conditions or resources in an ecosystem that keep a population at a certain size. Ecosystems only have a certain amount of food, water, space, and shelter to support a certain number of organisms. The relationship between numbers of organisms and the resources available in an ecosystem is often called the *balance of nature*. If any of the limiting factors change, animal and plant populations may also change, causing populations to increase or decrease in response to the change. Population increases may result in overcrowding. If plant populations increase, then populations of animals that consume that plant may also increase. If predator populations increase, then prey population will decrease. If prey population increases, then predator population will also increase because of increased availability of food. If water supply decreases, then the population of animals dependent on that water may decrease. If trees are removed or infected with diseases or parasites, then the population of animals dependent on the tree for food or shelter will also decrease. If organisms do not have enough space, then they will either have to move or will die. Space changes may caused by human influences or natural hazards.
- Vocabulary: *ecosystem*, *terrestrial*, *aquatic*, *biotic factors*, *populations*, *organisms*, *microorganisms*, *communities*, *abiotic factors*, *temperature*, *range of temperature*, *quantity of light*, *quantity of water*, *soil composition*, *limiting factors*, *food*, *space*, *shelter*, *introduced organism*, *overcrowding*, *availability*, *abundance*, *scarcity*, *disease*, *parasite*, *yield*, *independent variable*, *dependent*, *variable*, *constant*, *average*, *maximize*.

ELABORATION:

- Students will complete *Maximizing Garden Yield* student investigation guide.
- Based on the needs of the plant, students will monitor the abiotic and biotic factors that affect the plants growth over a multi-week period.
- Students will make observations and measurements of the plants growth using *Growth Data and Plant Type Data Tables* (attached).

****When observing measuring the growth, students should chose four or five plants to measure and observe each time they record their findings. Students may mark the plants chosen using Popsicle sticks.*

EVALUATION:

- Students will turn in the *Maximizing Garden Yield* student investigation guide.
- Teacher will observe reactions and interactions with other students and the plants.

Growth Data

Week	Plant A	Plant B	Plant C	Plant D
1	Height: # of Leaves: # of Fruits:	Height: # of Leaves: # of Fruits:	Height: # of Leaves: # of Fruits:	Height: # of Leaves: # of Fruits:
2	Height: # of Leaves: # of Fruits:	Height: # of Leaves: # of Fruits:	Height: # of Leaves: # of Fruits:	Height: # of Leaves: # of Fruits:
3	Height: # of Leaves: # of Fruits:	Height: # of Leaves: # of Fruits:	Height: # of Leaves: # of Fruits:	Height: # of Leaves: # of Fruits:
4	Height: # of Leaves: # of Fruits:	Height: # of Leaves: # of Fruits:	Height: # of Leaves: # of Fruits:	Height: # of Leaves: # of Fruits:
5	Height: # of Leaves: # of Fruits:	Height: # of Leaves: # of Fruits:	Height: # of Leaves: # of Fruits:	Height: # of Leaves: # of Fruits:
6	Height: # of Leaves: # of Fruits:	Height: # of Leaves: # of Fruits:	Height: # of Leaves: # of Fruits:	Height: # of Leaves: # of Fruits:

Plant Types Data

Plant Type	# of Days in Seed Stage	# of Days in Seedling Stage	# of Days in Mature Stage

Plant Type/Date of Observation	Length of Stem	Leaf Appearance	Leaf Width	Number of Leaves	Number of Buds

MAXIMIZING GARDEN YIELD CYCLES

Name

____

List five factors in each box.

Biotic Factors	*Abiotic Factors*	*Limiting Factors*

List the environmental factors which affect plant growth.

____

____

____

____

____

FIFTH GRADE LESSON PLAN
Building Biodiversity

SUBJECT / GRADE LEVEL: Fifth Grade -Life Science.

ACADEMIC STANDARDS AND PERFORMANCE INDICATORS:

Science:

5.L.4B.1 Analyze and interpret data to explain how organisms obtain energy and classify organisms as producers, consumers including herbivore, carnivore, and omnivore, or decomposers such as fungi and bacteria.

5.L.4B.2 Develop and use models of food chains and food webs to describe the flow of energy in an ecosystem.

LESSON OBJECTIVES:

TSW identify, sketch, label, measure, describe, and classify, organisms observed in the garden.

MATERIALS:

- *Biodiversity in Our Garden* handouts (attached)
- Clipboards
- Large, three-ring binder with dividers for *Producers, Consumers, and Decomposers*
- iPad or camera for photographing organisms
- Magnifiers hand lenses
- Illustration utensils
- Images of frequent garden visitors from around the garden
- Plastic bags for collecting samples
- Newspaper for pressing plants
- Chart paper
- Disposable newspapers and magazines
- Tape
- Scissors
- Blank labels or sticky notes
- Markers
- *Building Biodiversity* student investigation guide (attached)
- *Building Biodiversity* teacher guide (attached)

ENGAGEMENT:

- Teacher will pose the question, "**What is biodiversity and why is it important?**"

EXPLORATION:

- Students will explore the garden with their small groups for twenty minutes, trying to identify examples of producers, consumers, and decomposers.

EXPLANATION:

- Using the *Building Biodiversity* teacher guide, the teacher will guide students through the completion of the *Building Biodiversity* student investigation guide.

ELABORATION:

- Students will complete *Building Biodiversity* student investigation guide.

EVALUATION:

- Students will turn in the *Building Biodiversity* student investigation guide.
- Students will be assessed on accuracy and completion of their *Building Biodiversity* student investigation guides as well as their *Organism Classification Collages.*

RESOURCES:

1. *Biodiversity In Our Garden* handout (attached).
2. *Garden Food Web* student example (attached).

TEACHER GUIDE
BUILDING BIODIVERSITY

BACKGROUND:

Students should understand that there are a variety of living things in a garden that make up the food web.

ENGAGEMENT:

- Teacher will pose the question, "**What is biodiversity and why is it important?**"

EXPLORATION:

- Establish boundaries for students to explore and review safety procedures, specifically that some organisms may be poisonous or harmful. Students explore the garden with their small groups for twenty minutes, trying to identify examples of producers, consumers, and decomposers. Then they document evidence on their handouts, being specific as possible about details and context.
- *Organism Classification Collages*: Teacher labels a piece of chart paper or poster board with the name of a group of organisms. Students search for pictures of organisms in old magazines and newspapers. They cut out the picture and attach it to the appropriate chart or poster. Then they use a label or a sticky note to write a distinct characteristic of that group of organisms. Display the completed examples as *Organism Classification Collages* in the classroom. Refer to them or add to them as needed throughout the year.
- Once each student collects examples of organisms in each classification group, students will use the *Biodiversity in Our Garden* handouts to create a sheet for each organism. The handouts should be filed in a large three-ring binder with dividers labeled *Producers*, *Consumers*, and *Decomposers*.

EXPLANATION:

- Using the *Building Biodiversity* teacher guide, the teacher will guide students through the completion of the *Building Biodiversity* student investigation guide.
- Reassemble the class as one group and small groups share the organisms they discovered in the garden. Introduce and define biodiversity. Give examples. Return to the classroom and allow time for students to work in groups to complete their *Building Biodiversity* student investigation guides.

- Vocabulary: *food chain, food web, classify, producer, consumer, decomposer, energy, flowering, non-flowering, seeds, spores, cones, vertebrate, invertebrate, herbivore, carnivore, omnivore, fungi, bacteria, classification, catalogue, beneficial, harmful, structural adaptations, edible, inedible, locomotion, defense, camouflage, seed dispersal, food web, food chain, organism, biodiversity, sensory organs.*

ELABORATION:

- Students answer the second focus question: **How can we catalog all of the organisms we discover in the garden?** Teacher conducts informal observations to make sure students remain on task and follow safety procedures. Teacher initiates conferences with students to assess background knowledge and diagnose misconceptions.
- Teacher should encourage each group of students to complete a *Producer, Consumer, and Decomposer* matrix every two weeks throughout the school year.

EVALUATION:

- Students will turn in the *Building Biodiversity* student investigation guide.
- Students will be assessed on accuracy and completion of their *Building Biodiversity* student investigation guides as well as their *Organism Classification Collages.*

Biodiversity in Our Garden: Producers

Name:	
Date:	Time:
Sketch:	Description: Measurements:
Exact Location:	Crop or Not Crop?
Seeds or No Seeds?	Flowers or No Flowers?
Beneficial or Harmful?	Structural Adaptations:
Edible or Inedible?	Anatomy:
Other:	

Biodiversity in Our Garden: Decomposers

<table>
<tr><td colspan="2">Name:</td></tr>
<tr><td>Date:</td><td>Time:</td></tr>
<tr><td>Sketch:</td><td>Description:

Measurements:</td></tr>
<tr><td>Exact Location:</td><td>Beneficial or Harmful?</td></tr>
<tr><td>On what was it growing or feeding?</td><td>Fungi, Bacteria or Invertebrate?</td></tr>
<tr><td colspan="2">Structural Adaptations:</td></tr>
<tr><td colspan="2">Other:</td></tr>
</table>

Biodiversity in Our Garden: Consumers

Name:	
Date:	Time:
Sketch:	Description: Measurements:
Exact Location:	Beneficial or Harmful?
Invertebrate or Vertebrate?	What was it doing?
Herbivore, Carnivore, or Omnivore?	Structural Adaptations:
Other:	

BUILDING BIODIVERSITY

Name

Fill in the blocks to create a food web based only on organisms found in the garden. (Example: Sun > pepper plant > aphid > ladybug > wren > hawk – Do not use this example)

FIFTH GRADE LESSON PLAN
Compost Exploration

SUBJECT / GRADE LEVEL: Fifth Grade - Life Science.

ACADEMIC STANDARDS AND PERFORMANCE INDICATORS:

Science:

5.L.4A.1 Analyze and interpret data to summarize the abiotic factors including quantity of light and water, range of temperature, salinity, and soil composition of different terrestrial ecosystems and aquatic ecosystems.

5.L.4A.2 Obtain and communicate information to describe and compare the biotic factors including individual organisms, populations, and communities of different terrestrial and aquatic ecosystems.

LESSON OBJECTIVES: TSW analyze growth data to evaluate the effects of using varying amounts of compost to gain an understanding of the ideal soil composition for plant growth.

MATERIALS:

- Four pots with drainage holes
- Compost preferably made by students at school
- Bean or pea seeds
- Topsoil
- Pencils
- Measuring cups
- *Compost Exploration* student investigation guide (attached)
- *Compost Exploration* teacher guide (attached)

ENGAGEMENT:

- Teacher will pose the question, "**Can you feed dead plants to living plants?**"

EXPLORATION:

- Teacher will guide students through the setup of the compost experiment that will take place over a four week period of time.

EXPLANATION:

- Using the *Compost Exploration* teacher guide, the teacher will guide students through the completion of the *Compost Exploration* student investigation guide.

ELABORATION:

- Students will complete *Compost Exploration* student investigation guide.

EVALUATION:

- Students will turn in the *Compost Exploration* student investigation guide.
- Teacher will observe reactions and interactions with other students and the compost.

RESOURCES:

1. *Compost Exploration* handout (attached).

TEACHER GUIDE
COMPOST EXPLORATION

BACKGROUND:

Students should understand that compost adds vital organic matter to the soil to help garden plants thrive.

ENGAGEMENT:

- Teacher will pose the question, "**Can you feed dead plants to living plants?**"

EXPLORATION:

- **While we know plants benefit from compost, how much compost should be added to the soil to ensure the best plant growth?**
- Show students the materials and guide them through coming up with a procedure to test the question. Although the teacher has an idea of what procedure to use, students are more engaged when taking ownership in the planning process.
- Ask students which variables should be the same and which should be different. **Should we add the same amount of water to each pot? Why? Should we put all of the pots in the same location or a different location? Why?**
- Encourage students to revisit the original question if they get off track. We are testing the effects of adding compost, not the effects of light exposure and water.
- Guide students to identify the manipulated variable, the responding variable, and the controls for the investigation.
- Students will prepare the four pots following these steps or a variation of these steps based on what procedure they decide to follow.
  –Label the pots 1 to 4.
  –Fill pot 1 with 100 percent compost.
  –Using a measuring cup, measure part compost and an equal amount of topsoil. Mix the compost and topsoil together thoroughly. Fill pot 2 with this 1-to-1 compost-to-topsoil mixture.
  –Using a measuring cup, measure one part compost and three parts topsoil. Mix the compost and topsoil together thoroughly. Fill pot 3 with this 1-to-3 compost-to-topsoil mixture.
  –Fill pot 4 with 100 percent topsoil.
  –Plant three to four seeds of the same species in each pot.
  –Water your seeds following the instructions on the package. Make sure you add the same

amount of water to each pot.

- After four or five days, your seeds should have sprouted. Count the number of seeds that have sprouted in each of the four pots. Record the number of seeds that have sprouted. Thin the extra plants so you have one plant in each pot.
- Follow the growth of your plants for four to five weeks. Once a week, measure the height of the plant in each pot. Record the height of the plants. Note if any plants die.

EXPLANATION:

- Using the *Compost Exploration* teacher guide, the teacher will guide students through the completion of the *Compost Exploration* student investigation guide.
- While students have developed an understanding that humus produced from the composting process is beneficial to plant growth, they still need to understand that plants cannot rely solely on humus to grow.
- Plants need the right mixture of soils that allow them to get their basic needs. A good mixture holds and drains water at the same time, has plenty of nutrients, and has plenty of space to allow air flow so that the roots don't suffocate. The ideal soil for most "food" plants is loam. Loam is a mixture of the three main types of soil; sand, silt, and clay. For food production, it is helpful to add organic matter such as the humus created from composting. The compost conditions the soil to improve structure and texture allowing for a better retention of nutrients, water, and air.
- Vocabulary: *soil composition*, *nutrients*, *humus*, *decomposition*, *manipulated variable*, *responding variable*, *control*.

ELABORATION:

- Students will complete *Compost Exploration* student investigation guide three times per week for a total of four weeks.
- After four weeks, count the number of surviving plants. Compare the number of surviving plants in the different mixtures of compost and loam. Then measure the height of the plants once more.
- Allow students to illustrate or capture using photography, the growth of the four plants as they move through the different life cycle stages.

EVALUATION:

- Students will turn in the *Compost Exploration* student investigation guide.
- Teacher will observe reactions and interactions with other students and the compost.

COMPOST EXPLORATION

Name ____

Predict which plant will grow the best and explain your prediction.

____

____

____

____

____

Week 1

	Date	*Height*	*Date*	*Height*	*Date*	*Height*
Plant 1						
Plant 2						
Plant 3						
Plant 4						

FIFTH GRADE LESSON PLAN
Companion Planting

SUBJECT / GRADE LEVEL: Fifth Grade - Life Science.

ACADEMIC STANDARDS AND PERFORMANCE INDICATORS:

Science:

5.L.4B.3 Construct explanations for how organisms interact with each other in an ecosystem.

5.L.4B.4 Construct scientific arguments to explain how limiting factors or a newly introduced organism can affect an ecosystem.

LESSON OBJECTIVES:

TSW explore the significance of companion planting and investigate claims that some plants flourish when planted with another plant. TSW use quantitative and / or qualitative evidence to either support or refute claims.

MATERIALS:

- Seeds for companion plantings to add to the garden
- Measurement tools measuring tape and scales
- Trowel
- Compost or fertilizer
- Access to the internet or books on companion plants
- *Companion Planting* student investigation guide (attached)
- *Companion Planting* teacher guide (attached)

ENGAGEMENT:

Teacher will pose the question, "**What is a companion?**"

EXPLORATION:

- Teacher will introduce several companion planting examples.
- Students will research and decide a companion planting combination to test in the garden.

EXPLANATION:

- Using the *Companion Planting* teacher guide, the teacher will guide students through the completion of the *Companion Planting* student investigation guide.

ELABORATION:

- Once the plants have reached the adult stage of their life cycles, students should have enough data to construct an argument for or against the validity of companion planting claim. Their conclusions must provide evidence to support their arguments.

EVALUATION:

- Students will turn in the *Companion Planting* student investigation guide.
- Teacher will observe reactions and interactions with other students and the plants.

RESOURCES:

1. *Companion Planting* handout (attached).

TEACHER GUIDE
COMPANION PLANTING

BACKGROUND:

Students should explore the significance of companion planting and investigate claims that some plants flourish when planted with another plant. Students will use quantitative and / or qualitative evidence to either support or refute claims.

ENGAGEMENT:

- Teacher will pose the question, "**What is a companion?**"

EXPLORATION:

- Show students examples in the garden, or present students with pictures of plants that are taken over by other plants. Explain that some plants should not be planted near others because they compete for sunlight, water, nutrients, and space.
- Next, explain that many gardeners claim there are plants from different species that make great companions when planted next to each other.
- Discuss the meaning of the word *companion*, generate synonyms, and identify roots and origins.
- Explain that some gardeners claim that not only do companion plants grow well next to each other; there are some plants that actually grow better when planted next to their buddy plant.
- Instruct students to research companion plants and select a pair they are interested in testing. Some examples include:

  CARROTS AND CHIVES

  Chives are an excellent companion to carrots because they improve the flavor and growth of the carrots.

  BASIL AND TOMATOES

  The basil helps the tomatoes by improving the flavor and growth of the tomatoes and by repelling flies and mosquitoes.

  SQUASH AND BORAGE

  The borage helps squash by improving the flavor and growth. It also deters worms.

  CORN AND SOYBEANS

  Corn takes a lot of nitrogen from the soil. When soybeans die back, they add nitrogen back to

the soil. The soybeans also deter Japanese beetles and chinch bugs.
BROCCOLI AND ROSEMARY
Rosemary is a friend to broccoli because it helps to repel harmful insects.

- Post Research: Plants can be companions for many reasons. Planting them together may balance nutrient levels in the soil, repel harmful insects to one of the plants, improve flavors, improve the growth, and so on.
- Compile students findings for the whole class to view.
- Discuss the claims asking students to determine what evidence would support or refute the claims.
- If a claim suggests that one plant repels insects, students may suggest making observations to compare the number of harmful insects around the control plants that are planted separately to the number of insects that around the manipulated variable plants planted next to each other.
- If a claim suggests that planting two plants next to each other improves the growth of one of the plants, students may suggest taking growth measurements and comparing the measurements of the control plants that are planted separately to the manipulated variable plants that are planted next to each other. Besides measuring the height of the plants in both plots, students may also suggest measuring and comparing the weight of the harvested fruits from each plot.
- When testing a claim that suggests the growth of only ONE plant is improved, that should be planted as the control. For example, it is suggested that basil improves the growth of tomatoes, so you would plant tomato and basil plants next to each other as the manipulated variable. Then, plant only tomatoes alone as the control.

  ****There are opportunities to connect this to social studies curriculum such as techniques used by Native Americans planting the "three sisters" and techniques used in colonial gardens.*

EXPLANATION:

- Using the *Companion Planting* teacher guide, the teacher will guide students through the completion of the *Companion Planting* student investigation guide.
- Throughout time, gardeners have observed that some plants do better when grown near others. Arranging crops so they complement each other in some way is known as "companion planting". Plants that have different requirements for their basic needs make good garden buddies. These plants can be planted close together to save space since they will not compete for resources. Some combinations can actually help one or more of the "buddy" plants flourish, and companion planting is also practiced to prevent pest problems. Specific crops typically flowers and herbs are believed to repel certain pests, lure them elsewhere, or confuse them. Others entice beneficial insects, birds, and spiders that have an appetite for dining on problem visitors. Research supports some of the claims about which plants make good companions, but many are based on historical observation and tradition.

- Choose a test spot in the garden. To test the theory of companion planting, choose two separate plots or sections. Students should plant a set of companion plants together manipulated variable and plant a set without its companion control. Example: tomatoes and marigolds together, tomatoes alone.
- Students record the claim in science notebooks and make predictions about the outcome.
- Control all other variables that could impact growth such as soil quality, water, and sunlight availability. All of these variables should be kept consistent for both planting sites.
- Vocabulary: *life cycle*, *organism*, *habitat*, *ecosystem*, *beneficial*, *harmful*, *claim*, *evidence*, *control*, *manipulated variable*.

ELABORATION:

- Students will complete *Companion Planting* student investigation guide.
- Allow students to illustrate or capture using photography, the growth of the plants as they move through the different life cycle stages. Besides collecting the quantitative data, students may record qualitative observations in science notebooks and begin drawing conclusions based on their observations.
- Once the plants have reached the adult stage of their life cycles, students should have enough data to construct an argument for or against the validity of companion planting claim. Their conclusions must provide evidence to support their arguments.
- Students may also determine that there were variables that affected the results which make it impossible to support or refute any companion planting claims. These variables could be weather, human interference, etc...

EVALUATION:

- Students will turn in the *Companion Planting* student investigation guide.
- Teacher will observe reactions and interactions with other students and the plants.

COMPANION PLANTING

Name

____

	Plant #1	*Plant #2*	*Benefits*
Companion Planting #1			
Companion Planting #2			
Companion Planting #3			
Companion Planting #4			

FIFTH GRADE LESSON PLAN
Lettuce Eat Our Harvest

SUBJECT / GRADE LEVEL: Fifth Grade - Life Science.

ACADEMIC STANDARDS AND PERFORMANCE INDICATORS:

Science:

5.S.1B.1 Construct devices or design solutions to solve specific problems or needs: 1.) ask questions to identify problems or needs, 2.) ask questions about the criteria and constraints of the devices or solutions, 3.) generate and communicate ideas for possible devices or solutions, 4.) build and test devices or solutions, 5.) determine if the devices or solutions solved the problem and refine the design if needed, and 6.) communicate the results.

LESSON OBJECTIVES:

TSW work with a team to compete for the creation of the best recipe using produce harvested from the garden. TSW use appropriate safety procedures when preparing their recipes and select appropriate measurement tools / units in preparing and recording their recipe.

MATERIALS:

- Kitchen tools including measuring tools such as tablespoon, teaspoon, cups
- Harvest from the garden
- Cookbooks
- Kitchen tool safety procedures students/teachers generate
- Serving utensils/plates
- *Lettuce Eat Our Harvest* team recipe sheet
- *Lettuce Eat Our Harvest* teacher guide (attached)

ENGAGEMENT:

- Teacher will pose the question, "**Who has eaten food that they grew from a seed?**"

EXPLORATION:

- Teacher will prepare the class for the harvest recipe challenge.

EXPLANATION:

- Using the *Lettuce Eat Our Harvest* teacher guide, the teacher will guide students through the completion of the *Lettuce Eat Our Harvest* team recipe sheet.

ELABORATION:

- Students will complete the *Lettuce Eat Our Harvest* team recipe sheet.

EVALUATION:

- Students will turn in the *Lettuce Eat Our Harvest* team recipe sheet.
- Teacher will observe reactions and interactions with other students and the harvest.

RESOURCES:

1. *Lettuce Eat Our Harvest* handout (attached).

TEACHER GUIDE

LETTUCE EAT OUR HARVEST

BACKGROUND:

Students should try the produce that they have grown in the garden.

ENGAGEMENT:

- Teacher will pose the question, "**Who has eaten food that they have grown themselves from a seed?**"

EXPLORATION:

Eating what you've grown is the ultimate garden reward. Students love to be able to wash, chop, measure, stir and then eat what they've grown. Eating their harvest also promotes healthy eating practices and often allows students the opportunity to try foods they've never tried before. Though cooking with kids can seem daunting at first, using these simple rules will alleviate a lot of stress and keep kids safe.

- This lesson may take place over several days or in one lesson depending on how in depth the teacher wants to go. Students can research recipes, do taste testing, create a class cookbook or the class can simply create one recipe together using what is currently ready to harvest.
- Do cooking tasks one at a time so it's easier to keep track of what all students are doing. Separate activities into distinct time frames: harvest, wash hands and produce, chop / grate / tear, cook, then eat. Breaking students into small groups with an adult is ideal.
- Wash hands with soap and hot water. If you touch your hair, rub your eyes, go to the bathroom, etc. you need to wash your hands again. Make sure that kids are just as aware of this as you are. Roll up sleeves and pull back hair.
- Wash any produce that you re going to use, and remove bad spots.
- Teachers may choose to do the cooking activities themselves or simply challenge students to create recipes that don't require cooking like salsa, sandwiches, cold soups, salads, etc...

EXPLANATION:

- **Look around the garden, what looks ready to harvest? What should we do with all of this harvest?** Students will choose one or two items from the garden to be their main ingredient teachers may want students to research recipes that include items currently growing before choosing ingredients.

- Introduce the challenge: **Which group can create the tastiest recipe using ingredients from the garden that are ready to be harvested?** Teams will create a recipe.
- Other Challenge options: **Who can create the healthiest recipe? The simplest recipe? The least expensive recipe?**
- Students can research nutritional information or average pricing for ingredients to include with their recipe.
- Criteria: The teacher should set certain parameters for student groups must follow to be considered as the winner of the challenge. These requirements may include:
  –The number of ingredients or source of ingredients (does teacher or student have to bring in additional items?).
  –The tools available (is a heat source needed?).
  –A time limit to challenge team work and leadership skills.
  –A precise list of measurement quantities used so that the recipe can be replicated.
- Discuss possibilities for using the main harvested ingredient. Pose questions such as: **What kind of food do you like to eat? How can we use this ingredient in a recipe that others will enjoy eating?**
- Student teams will present recipes to the class and the teacher or class will vote on the recipe or recipes which will be prepared for the class to eat.
- Each team will complete a *Lettuce Eat Our Harvest* recipe sheet.
- The teacher should lead the class in creating a list of safety rules and procedures before passing out any tools. It may be necessary to introduce students to kitchen tools. Relate items such as cups, teaspoons, etc. to math lessons. Explain the importance of measuring and recording measurements while they are "cooking" so that the recipe can be reproduced correctly.
- Once the safety rules and procedures have been established, the recipe preparation can begin.

ELABORATION:

- Create a class "cookbook" with recipes that the class has created using garden ingredients over the school year.
- Ask the cafeteria staff to recreate the winning recipe to share with the grade level or school.
- Host a parent's lunch with food made by the students using garden harvest.
- Write fractions for the recipe as a whole.
- Double or triple the recipe to serve more people.

EVALUATION:

- Teams will turn in the *Lettuce Eat Our Harvest* team recipe sheet.
- Teacher will observe reactions and interactions with other students and the harvest.

<u>LETTUCE EAT OUR HARVEST</u>

Team Members

____

____

Recipe Title:

____

List the ingredients in your recipe. Circle the items that will be harvested from the garden.

Ingredients

List the steps necessary to prepare your recipe.

____

____

____

____

____

FIFTH GRADE LESSON PLAN
Digging Into Decomposition

SUBJECT / GRADE LEVEL: Fifth Grade - Life Science.

ACADEMIC STANDARDS AND PERFORMANCE INDICATORS:

Science:

5.L.4A.1 Analyze and interpret data to summarize the abiotic factors including quantity of light and water, range of temperature, salinity, and soil composition of different terrestrial ecosystems and aquatic ecosystems.

5.L.4A.2 Obtain and communicate information to describe and compare the biotic factors including individual organisms, populations, and communities of different terrestrial and aquatic ecosystems.

5.L.4B.1 Analyze and interpret data to explain how organisms obtain energy and classify organisms as producers, consumers including herbivore, carnivore, and omnivore, or decomposers such as fungi and bacteria.

5.L.4B.2 Develop and use models of food chains and food webs to describe the flow of energy in an ecosystem.

5.L.4B.4 Construct scientific arguments to explain how limiting factors or a newly introduced organism can affect an ecosystem.

LESSON OBJECTIVE:

TSW closely observe the process of decomposition and analyze the environmental factors that affect the rate of decomposition.

MATERIALS:

- Two slices of bread for each group
- One lemon sliced in quarters for each group
- One banana sliced in thirds for each group
- Six plastic bags for each group
- Microscopes (if available)
- Hand lenses
- *Digging Into Decomposition* student investigation guide (attached)
- *Digging Into Decomposition* teacher guide (attached)

ENGAGEMENT:

- Teacher will pose the question, "**What turns compost into compost?**"

EXPLORATION:

- Students will investigate the environmental factors which affect the rate of decomposition.

EXPLANATION:

- Using the *Digging Into Decomposition* teacher guide, the teacher will guide students through the completion of the *Digging Into Decomposition* student investigation guide.

ELABORATION:

- Students will complete *Digging Into Decomposition* student investigation guide.

EVALUATION:

- Students will turn in the *Digging Into Decomposition* student investigation guide.
- Teacher will observe reactions and interactions with other students and the decomposing items.

RESOURCES:

1. *Digging into Decomposition* handout (attached).

TEACHER GUIDE
DIGGING INTO DECOMPOSITION

BACKGROUND:

Students should understand that decomposition creates compost and the decomposition rate is dependent on several environmental factors.

ENGAGEMENT:

Teacher will pose the question, "**What turns compost into compost?**"

EXPLORATION:

- Review parts of a food chain sun, producer, consumer, and decomposer. Discuss what takes place if something happens to one of the links.
- Next, explain that the first consumers in a compost pile food web are *microorganisms*. *Bacteria* and *fungi* are the foundation of the compost pile food web and, without them, a compost pile food web would collapse.
- Finally, explain that many environmental factors affect how slow or fast rate bacteria and fungi consume organic matter.
- **What environmental factors affect the rate speed of decomposition?**

EXPLANATION:

- Using the *Digging Into Decomposition* teacher guide, the teacher will guide students through the completion of the *Digging Into Decomposition* student investigation guide.
- Students will help the teacher list environmental factors: temperature, water, space, light, etc...
- Explain that the class will be examining the following claim: high moisture content and warm temperatures encourage rapid growth of fungi and bacteria. Both of these factors help fungi and bacteria thrive. Explain that students will develop a procedure to test this claim and use evidence to support or refute the claim.
- Pass out the material to each group two pieces of bread, one orange or lemon sliced into four segments, one banana sliced in thirds, six plastic bags and a permanent marker.
- Allow students time to discuss how they will test the claim using these materials with their group. Students may suggest testing the decomposition of the food in two separate environments. Guiding questions: **What will we need to keep the foods moist?** Water. **What**

is the opposite of lots of moisture? Dry. **Where can we put the foods so they will stay warm?** Outside. **What environment would be the opposite of this?** Refrigerator.

- Allow students time to record a procedure asking them to use steps such as first, next, then, after one week, etc... Next, allow students to read and compare procedures from other groups.
- As a class, determine what procedure is best.

****Prior to setting up the investigation with students, you may want to rub the oranges / lemon slices on the floor to pick up bacteria and then let them sit out overnight.*

- Vocabulary: decomposition, fungi, bacteria, invertebrates, consumers, environmental factors, rate, producers, consumers, decomposers.

ELABORATION:

- Students will complete *Digging Into Decomposition* student investigation guide.
- After one week, ask students if they can determine where more fungi and bacteria are present. Allow students to use microscopes and hand lenses to try and view the microbes.

EVALUATION:

- Students will turn in the *Digging Into Decomposition* student investigation guide.
- Teacher will observe reactions and interactions with other students and the decomposing items.

DIGGING INTO DECOMPOSITION

Name

____

Complete three observations each day.

	Bread	*Lemon*	*Banana*
Day 1 observation			
Day 2 observation			
Day 3 observation			

FIFTH GRADE LESSON PLAN
Worm Bin Builders

SUBJECT / GRADE LEVEL: Fifth Grade - Life Science.

ACADEMIC STANDARDS AND PERFORMANCE INDICATORS:

Science:

5.L.4A.1 Analyze and interpret data to summarize the abiotic factors including quantity of light and water, range of temperature, salinity, and soil composition of different terrestrial ecosystems and aquatic ecosystems.

5.L.4A.2 Obtain and communicate information to describe and compare the biotic factors including individual organisms, populations, and communities of different terrestrial and aquatic ecosystems.

5.L.4B.1 Analyze and interpret data to explain how organisms obtain energy and classify organisms as producers, consumers including herbivore, carnivore, and omnivore , or decomposers such as fungi and bacteria.

5.L.4B.2 Develop and use models of food chains and food webs to describe the flow of energy in an ecosystem.

5.L.4B.4 Construct scientific arguments to explain how limiting factors or a newly introduced organism can affect an ecosystem.

LESSON OBJECTIVES:

TSW gain an understanding of why humans compost, the role worms play in decomposition, the importance of nutrient rich humus, and the environment characteristics that support a variety of organisms in a compost bin.

ENGAGEMENT:

- Teacher will pose the question, "**What kind of house does a worm like to live in?**"

EXPLORATION:

- The teacher will explain that *composting* is the process of organic materials such as fallen leaves and left over food scraps decomposing into nutrient rich humus.
- The teacher will also explain benefits of composting.

MATERIALS:

- ½ to 1 pound of Red Wiggler Worms
- Soil
- Trowel
- Newspaper
- Water
- Food scraps
- Scale
- Thermometer
- 10 gallon plastic bin. Drill ⅛ inch holes in the bottom for drainage and 1/16 inch holes around the top of the bin and on the lid to provide ventilation. Add a tray underneath the bin to catch any drainage or set it down in a second 10 gallon bin
- *Worm Bin Builders* student investigation guide (attached)
- *Worm Bin Builders* teacher guide (attached)

EXPLANATION:

- Using the *Worm Bin Builders* teacher guide, the teacher will guide students through the completion of the *Worm Bin Builders* student investigation guide.

ELABORATION:

- While maintaining the worm bin, students may do many explorations or assignments that allow them to explain the understanding of the concept under examination.

EVALUATION:

- Students will turn in the *Worm Bin Builders* student investigation guide.
- Teacher will observe reactions and interactions with other students and the worms.

RESOURCES:

1. *Worm Bin Builders* handout (attached).

TEACHER GUIDE
WORM BIN BUILDERS

BACKGROUND:

Students should understand the role worms play in decomposition, the importance of nutrient rich humus, and the environment characteristics that support a variety of organisms in a compost bin.

ENGAGEMENT:

- Teacher will pose the question, "**What kind of house does a worm like to live in?**"

EXPLORATION:

- The teacher will ask students to recall what plants need in order to survive and guide students through a discussion about how plants depend on nutrients.
- Next, the teacher will explain that they will be exploring the process of composting which has multiple benefits to the environment.
- The teacher will explain that composting is the process of organic materials such as fallen leaves and left over food scraps decomposing into nutrient rich humus. The teacher will also explain benefits of composting.

BENEFITS:

–We use nonrenewable resources to produce the energy it takes to transport materials to landfills. Landfills are generally far away from humans. Since landfills are far away, we use more resources to transport waste to these landfills. Composting reduces the amount of energy used to transport these materials.

–Reducing the amount of waste also benefits the landfill space issue. While the human population on earth continues to increase, the amount of human waste also increases. However, the amount of space for landfills decreases. Composting decreases the amount of waste sent to landfills, therefore decreasing the amount of space used to store the waste.

–Instead of buying fertilizers, gardeners and farmers can use the humus created from composting to provide soil with nutrients that plants require for healthy growth. Adding the humus enables the existing soil to better retain nutrients, moisture, and air. Composting is cost effective and helps the environment by using a natural process to improve new growth.

- Next, the teacher will explain that the class will be observing the process of decomposition by adding worms to the compost bin they will create and maintain. The process of using worms to compost is called *vermicomposting*.

- The teacher may provide or ask students to research the environmental characteristics worms need to survive.

EXPLANATION:

- Students will complete *Worm Bin Builders* student investigation guide.
- Teacher will lead students in the building of the worm bin. The first step is preparing the bedding. Demonstrate how to prepare the worm bin bedding by modeling how to tear newspaper into ½" to 1" wide strips. Allow students to tear about ten sheets of newspaper lengthwise with the grain. Dunk the newspaper strips in water the paper should be as wet as a wrung-out sponge/not dripping wet, and add to the bin.
- Fluff the "bedding" like a big salad so that there is air between the papers and fill the plastic bin until it is half full.
- Sprinkle about a quart of soil over the top.
- Students draw a diagram of the worm bin from an overhead view. Students can divide the bin into sections and number the sections. Students may use this diagram for placing food waste in different sections of the bin. This will enable students to keep track of where and when the food waste is placed.
- With a trowel or your hand, dig halfway through the bedding in the center of the box.
- Tuck your red wigglers into their nice, moist bed and then cover them with the bedding.
- Before feeding the worms, guide students through using the record sheet to record the date, weight of food, water added, burying location, type of food, and the temperature during research, students may have discovered that a worm s environment must be moist and warm, therefore it is important that they keep a log of the temperature of the bin to reference later.
- Dig a hole in the bedding material in a corner section of the plastic box. Place a small amount of cafeteria scraps in the hole, cover it with bedding, and record the food location or mark with Popsicle sticks.
- Worms like any fruit, vegetable, or grain/bread. They also like coffee grounds and filters, tea bags, and crushed eggshells. Stay away from feeding them meat and dairy.

ONGOING MAINTENANCE:

- Let the worm bin rest by not adding any additional food for about a week. This allows the worms a chance to get used to their new environment and for the food to begin to decompose.
- Ask students to observe how long it takes for this initial food to decompose. Have them check every few days.
- Feed the worms every two to four days, always burying the food under the bedding to avoid fruit flies or other pests. Do not overfeed.
- Worms adapt easily to living on plant waste inside a worm bin. To keep your worms healthy, make sure the worm bin is damp not wet and keep it in a cool spot out of direct sunlight.

Ideal temperatures for the worms range from 55° to 77° degrees. Students will have the task of ensuring that the worms have everything they need to survive in the bin darkness, moisture, moderate temperature, food and bedding. Add more bedding as needed.

- Once all the bedding is gone and your compost bin smells like a fresh forest usually after 3–5 months it's time to harvest. It's better to harvest too early than too late, which can kill your worms. Any bits of food left over can be put back into the next worm bin or added to an outside compost bin.

FAST AND MESSY HARVESTING METHOD:

- To separate the worms from the compost, empty the contents of the worm compost bin onto a tarp or old plastic tablecloth. Worms do not like light and will wiggle into the pile. Wait a few minutes. Then with your trowel or your hands, remove the top layer of the compost pile until the worms are visible.
- Then wait again, be patient, and continue removing the compost. Repeat until all the worms are found. All the worms can go into an outside compost bin or can start another compost box.

ELABORATION:

- Following a worm observation, ask students to interpret the meaning of the following Charles Darwin quote:
  "It may be doubted whether there are many other animals which have played so important a part in the history of the world as these lowly organized creatures."
  —Charles Darwin on Earthworms 1881
- While maintaining the bin, students may do many explorations or assignments that allow them to explain the understanding of the concept under examination.
- Students may write a paper describing what happens when decomposers break down organic material, explaining how the material is used to create new life. Students may also illustrate a diagram of the process. Fallen fruit / leaves decomposers such as worms, mites, bacteria, fungus to new plant growth.
- After many weeks, use hand lenses to closely observe the material from the compost bin. Pour the material on newspaper and provide students Popsicle sticks to move around the material.
- Ask students questions like: **Can you see food? What kinds? Are there any new compost critters? Are the worms active? Is there a type of food that the worms decompose at faster rates? Are the worms getting the energy they need from their environment? What evidence shows this?**
- Students may suggest conducting investigations to answer their own inquiries. For example, if a student wonders if worms really prefer damp environments, they may put a few worms in between wet and dry soil to observe which direction the worms will move.

- If some of the worms are not surviving, students may use data to construct arguments for what biotic or abiotic factors are affecting the survival of the worms. Next, students may develop methods to solve the problem.
- Students could research the role of other decomposing organisms that exist in a compost food web.

EVALUATION:

- Students will turn in the *Worm Bin Builders* student investigation guide.
- Teacher will observe reactions and interactions with other students and the worms.

WORM BIN BUILDERS

Name

___

What is vermicomposting?

___

___

___

List three benefits of composting.

___

___

___

List three foods that work well when vermicomposting.

___

___

___

List three foods that should be avoided when vermicomposting.

___

___

___

FIFTH GRADE LESSON PLAN
What Does It Weigh?

SUBJECT / GRADE LEVEL: Fifth Grade - Math.

ACADEMIC STANDARDS AND PERFORMANCE INDICATORS:

Math:

5.MDA.1 Convert measurements within a single system of measurement: customary i.e., in., ft., yd., oz., lb., sec., min., hr. or metric i.e., mm, cm, m, km, g, kg, mL, L from a larger to a smaller unit and a smaller to a larger unit.

LESSON OBJECTIVE:

TSW apply their knowledge of measurement units and tools while harvesting produce.

MATERIALS:

- Large plastic bins for each group to use for harvesting
- Scale
- *What Does It Weigh?* student investigation guide (attached)
- *What Does It Weigh?* teacher guide (attached)

ENGAGEMENT:

- Teacher will pose the question, "**How do you know how much produce you've harvested?**"

EXPLORATION:

- Teacher will explain the procedures for harvesting produce from the garden.
- Students will be divided into teams for harvesting.
- Each team should estimate the amount of produce that their team will harvest.

EXPLANATION:

- Using the *What Does It Weigh?* teacher guide, the teacher will guide students through the completion of the *What Does It Weigh?* student investigation guide.

ELABORATION:

- Students will complete *What Does It Weigh?* student investigation guide.

EVALUATION:

- Students will turn in the *What Does It Weigh?* student investigation guide.
- Teacher will observe reactions and interactions with other students and the produce.

RESOURCES:

1. *What Does It Weigh?* handout (attached).

TEACHER GUIDE
WHAT DOES IT WEIGH?

BACKGROUND:

Students should understand that math is an important skill to use when harvesting produce.

ENGAGEMENT:

- Teacher will pose the question, "**How do you know how much produce you've harvested?**"

EXPLORATION:

- Teacher will explain the procedures for harvesting produce from the garden. Students will be divided into teams for harvesting. Each team should estimate the amount of produce that their team will harvest.

EXPLANATION:

- Using the *What Does It Weigh?* teacher guide, the teacher will guide students through the completion of the *What Does It Weigh?* student investigation guide.
- Call time and gather students back together. Allow students to lift the bins and estimate which they feel is heaviest. Also, revisit initial recorded estimates.
- Allow students to brainstorm / discuss how they can find the weight of the produce without the weight of the plastic bins students may suggest finding the weight of an empty bin and subtracting that weight from the weight of the bin filled with produce.
- Once the produce has been weighed and the weight of the bins is subtracted, students can interpret the data.
- Possible guiding questions:
  –Which team harvested the most produce?
  –How many more pounds of produce did team A pull than team B?
  –How many pounds of produce did we harvest as a class?
  –Which team estimate was closest to the actual weight of produce harvested?
  –What is the weight of the weeds your team pulled converted to ounces?
  –What is the weight of the weeds our whole class pulled converted to ounces?

ELABORATION:

- Students will complete *What Does It Weigh?* student investigation guide.

EVALUATION:

- Students will turn in the *What Does It Weigh?* student investigation guide.
- Teacher will observe reactions and interactions with other students and the produce.

WHAT DOES IT WEIGH?

Name

____

Your Harvest Estimated Weight	
Your Actual Harvested Weight	
Total Class Harvested Weight	
Your Weight Converted to Ounces	
Class Weight Converted to Ounces	

Describe the method that you used to harvest produce from the garden?

____

____

____

What type of produce did you harvest from the garden?

____

____

____

FIFTH GRADE LESSON PLAN
Soil Composition

SUBJECT / GRADE LEVEL: Fifth Grade - Life Science.

ACADEMIC STANDARDS AND PERFORMANCE INDICATORS:

Science:

5.L.4A.1 Analyze and interpret data to summarize the abiotic factors including quantity of light and water, range of temperature, salinity, and soil composition of different terrestrial ecosystems and aquatic ecosystems.

LESSON OBJECTIVE:

TSW investigate soil samples in order to visualize different types of soil and become familiar with their physical properties and composition.

MATERIALS:

- Trowel
- Tray to collect soil sample
- Soil sample
- Water
- Clear jar or bottle
- *Soil Composition* student investigation guide (attached)
- *Soil Composition* teacher guide (attached)

ENGAGEMENT:

- Teacher will pose the question, "**What would we find if we separated all of the things that make up soil?**"

EXPLORATION:

- Teacher will facilitate the setup of the soil experiment.

EXPLANATION:

- Using the *Soil Composition* teacher guide, the teacher will guide students through the completion of the *Soil Composition* student investigation guide.

ELABORATION:

- Students will complete *Soil Composition* student investigation guide.

EVALUATION:

- Students will turn in the *Soil Composition* student investigation guide.
- Teacher will observe reactions and interactions with other students and the soil.

RESOURCES:

1. *Soil Composition* handout (attached).

TEACHER GUIDE

SOIL COMPOSITION

BACKGROUND:

Students should understand that soil is composed of both organic and inorganic components.

ENGAGEMENT:

- Teacher will pose the question, "**What would we find if we separated all of the things that make up soil?**"

EXPLORATION:

- Preparing the soil layers:
  –Take several soil samples from around the school grounds.
  –Add the soil to a large jar or bottle.
  –While students watch, pour water into the tube and ask them to consider why there are air bubbles. Explain that soil is also made up of air and water.
- Put a lid on the bottle or jar and shake vigorously. Explain that this process separates the different types of soil from the sample. The largest soil particles settle first on the bottom.
- Allow the sample to settle overnight.

EXPLANATION:

- In this investigation, the teacher and students collect a soil samples from around the school, add it to water, shake the ingredients, and allow them to settle for observation of the different types of soil particles.
- Soil type refers to the different sizes of mineral particles in a particular sample. Soil is made up in part of finely ground rock particles, grouped according to size. Soil types include sand, silt, clay. In addition to mineral matter, soil consists of air, water, and organic material such as decomposed plants and animals.
- The following day, **ask students to illustrate the settled soil**.
- Instruct students to label the different types of soil based on the descriptions you provide it would be helpful to illustrate zoomed in particle sizes of soil on the board while describing how the soil types settled so students can visualize the different particles: *sand*, *silt* and *clay*.
- **Ask students why some particles or grains are on the bottom.** Compare the particles introducing the words *coarse* and *fine*.

- Tell students that sand is the coarsest and has the largest particle size so it normally settles first.
- Next, tell them that clay is the finest particle and has the smallest particle so it normally settles last.
- Explain that silt is finer than sand, but coarser than clay. Have the students guess where the silt would fall.
- Finally, **ask students if they see anything floating at the top of their water** organic matter and humus.
- Introduce *humus* as once living organisms such as leaves and animals.
- Students will complete *Soil Composition* student investigation guide.
- Vocabulary: *particle*, *physical properties*, *texture*, *sample*, *clay*, *silt*, *sand*, *humus*, *loam*, *decomposition*, *coarse*, *fine*.

ELABORATION:

- Students may study the different types of soil further. Get samples of each soil type and observe each type closely with hand lenses. Students will record the properties of each type. After observation of each type, students may:
  –Experiment with how each type reacts to or holds water.
  –Compare two types on a Venn Diagram.
  –Develop soil "raps".

EVALUATION:

- Students will turn in the *Soil Composition* student investigation guide.
- Teacher will observe reactions and interactions with other students and the soil.

SOIL COMPOSITION

Name

____

What would we find if we separated all of the things that make up soil?

____

____

____

____

List eight components that we might find in soil by writing them in the pieces of the pie chart below.

FIFTH GRADE LESSON PLAN

Final Scavenger Hunt

SUBJECT / GRADE LEVEL: Fifth Grade - Inquiry and Life Science.

ACADEMIC STANDARDS AND PERFORMANCE INDICATORS:

Science:

5.L.4A.1 Analyze and interpret data to summarize the abiotic factors including quantity of light and water, range of temperature, salinity, and soil composition of different terrestrial ecosystems and aquatic ecosystems.

5.L.4B.2 Obtain and communicate information to describe and compare the biotic factors including individual organisms, population, and communities of different terrestrial and aquatic ecosystems.

5.L.4B.4 Construct scientific arguments to explain how limiting factors including food, water, space, and shelter or a newly introduced organism can affect an ecosystem.

5.S.1A.3 Plan and conduct controlled scientific investigations to answer questions, test hypotheses and predictions, and develop explanations: 1.) formulate scientific questions and testable hypotheses, 2.) identify materials, procedures, and variables, 3.) select and use appropriate tools or instruments to collect qualitative and quantitative data, and 4.) record and represent data in an appropriate form. Use appropriate safety procedures.

LESSON OBJECTIVES:

TSW identify, sketch, label, measure, describe, and classify, organisms observed in the garden. TSW explore the significance of companion planting and investigate claims that some plants flourish when planted with another plant. TSW use quantitative and / or qualitative evidence to either support or refute claims. TSW analyze growth data to evaluate the effects of using varying amounts of compost to gain an understanding of the ideal soil composition for plant growth. TSW closely observe the process of decomposition and analyze the environmental factors that affect the rate of decomposition. TSW identify and describe the biotic, abiotic, and limiting factors of the garden ecosystem and classify and compare them according to their characteristics. TSW analyze and interpret data about the abiotic factors present in the garden to summarize the availability of resources necessary for optimum plant growth. TSW investigate soil samples in order to visualize different types of soil and become familiar with their physical properties and composition. TSW gain an understanding of why humans compost, the role worms play in decomposition, the importance of nutrient rich humus, and the environment characteristics that support a variety of organisms in a compost bin.

MATERIALS:

- *Building Biodiversity* student investigation guide
- *Companion Planting* student investigation guide
- *Compost Exploration* student investigation guide
- *Digging Into Decomposition* student investigation guide
- *Maximizing Garden Yield* student investigation guide
- *Soil Composition* student investigation guide
- *Worm Bin Builders* student investigation guide
- Paper lunch sacks to collect items

ENGAGEMENT:

- Teacher will pose the question, "**What is a term that you've learned during our garden lessons?**"

EXPLORATION:

- Teacher will poll students for plant terms until all of the terms for the scavenger hunt have been discussed.

EXPLANATION:

- Using the *Scavenger Hunt* chart, the teacher will guide students through the explanation of all terms in the chart.
- After reviewing the *Scavenger Hunt* chart, students will go to the garden in order to find the items.

ELABORATION:

- Students can compost all of the items from the scavenger hunt including the chart and the paper bag if there is nowhere to compost the items, this could also be turned into a digital scavenger hunt where students simply take pictures of the items in the chart.

EVALUATION: IN THE CLASSROOM

- Students will show teacher items in the bag as the teacher checks them off of the *Scavenger Hunt* chart.

RESOURCES:

1. *Scavenger Hunt* handout (attached).

SCAVENGER HUNT

Name

____

Instructions: Find all of the items listed below:

Item	Where did you find it?
1) A producer	
2) A consumer	
3) A decomposer	
4) Humus	
5) Abiotic part of the ecosystem	
6) Biotic part of the ecosystem	
7) Sand particles	
8) Clay particles	
9) Vermicompost	
10) Compost	

LESSONS FOR SIXTH GRADE

SIXTH GRADE LESSON PLAN
Powerful Pollinators

SUBJECT / GRADE LEVEL: 6th Grade - Science - Duration: Two sixty minute lessons.

ACADEMIC STANDARDS AND PERFORMANCE INDICATORS:

Science:

6.L.5B.3 Develop and use models to compare structural adaptations and processes that flowering plants use for defense, survival and reproduction.

LESSON OBJECTIVES:

TSW identify the female and male reproductive parts of a flowering plant and draw and explain the function of those parts. TSW will discover the specialized relationship between plants and their pollinators.

MATERIALS:

- *Design a Flower* Part I and II student worksheets (attached)
- *Powerful Pollinator* teacher guide (attached)
- *Powerful Pollinator* Teacher's Rubric (attached)

ENGAGEMENT:

- Students will write three words to describe the fragrance of a flower. Teacher will write *stinky, smelly* and *gross*.

EXPLORATION:

- DAY 1: The teacher will begin by writing the words stinky, smelly and gross to describe the fragrance of a flower.
- DAY 2: Students will share their findings on specialized pollinators.

EXPLANATION:

- DAY 1: The teacher will follow up the engagement question with the engaging video titled *The Corpse Flower: Behind the Stink* by National Geographic: https://www.youtube.com/watch?v=l-uK5ulLBr8.

- Using the *Powerful Pollinator* teacher guide, the teacher will guide students through the completion of Part I of *Design a Flower.* Teacher will encourage students to explore the different types of specialized pollinator / flower relationships for homework to be shared with the class on the following day.
- DAY 2: Students will share their pollinator homework findings. Teacher can share additional specialized pollinators such as the Yucca moth and the Melipona bee.

EVALUATION:

- DAY 1: Students will turn in Part I of *Design a Flower.*
- DAY 2: Students will turn in Part II of *Design a Flower.*

ELABORATION:

- An extension to this lesson would be planting things such as sunflowers to attract pollinators to the school garden.

RESOURCES:

1. *The Corpse Flower: Behind the Stink* by National Geographic: https://www.youtube.com/watch?v=l-uK5ulLBr8.
2. *Design a Flower* Part I and II worksheets (attached).
3. *Powerful Pollinator* teacher guide and rubric (attached).

TEACHER GUIDE
POWERFUL POLLINATOR

BACKGROUND:

Students should understand that the purpose of a flower is to enable a plant to reproduce through pollination and formation of seeds.

ENGAGEMENT:

- Ask students to list four parts of a car and four part of a flower. Students will likely have more trouble with the flower parts. Explain that:

  1. Plants are responsible for the food that enables us to live. The flower parts make the production of most food possible. Therefore, the parts of a flower are literally a matter of life and death.

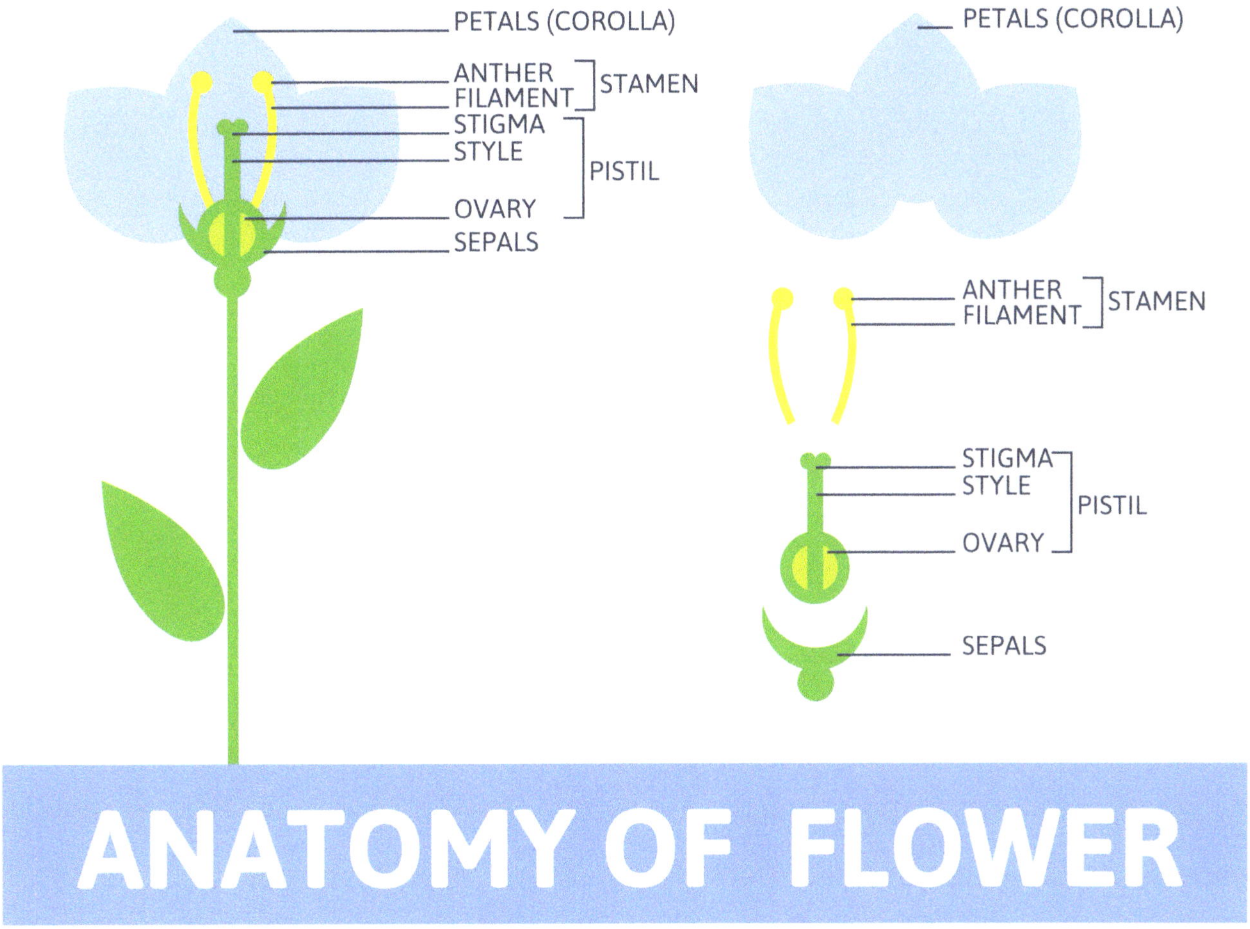

EXPLORATION:

Guide students through the completion of Part I of *Design a Flower* using the drawing above as well as the descriptions below:

- *Stamen*: the male flower part. It is made up of the filament stalk that holds up the anther. Pollen is contained in the anther and it contains the male sperm.
- *Pistil*: the female part of the flower. It has a sticky tissue at the tip called the stigma. It is sticky in order to receive the pollen. Below the sigma is a tube that called the style. It is designed to carry pollen from the stigma to the ovary. The ovary contains eggs, which will then be fertilized by the sperm in the pollen.
- *Sepals*: leaf-like structures located beneath the petals. They support the flower petals as well as protect the flower bud before it opens. The sepals are usually green.
- *Petals*: often colorful leaf-like structures that attract animals, insects and other pollinators.

SEXUAL REPRODUCTION:

- Sexual reproduction involves the fertilization of eggs by male sperm contained in the pollen. Once fertilized the eggs or ovules mature into seeds.
- Therefore, when you buy a pack of seeds you are essentially buying a pack of fertilized plant eggs that are sleeping until you wake them up.
- In future lessons, you will learn how seeds manage to travel to the perfect destination to wake them up.

Design a Flower Name ______________________________

Part I: Write the flower parts in the appropriate location on the drawing below. In addition, write the function of each flower part.

- **Sepals-**

- **Petals-**

- **Pistil-**

- **Stamen-**

Design a Flower

Name______________________

Part II: Design a flower that contains the four parts studied in our lesson. Label each part appropriately and write at least one sentence describing what each part does on your unique flower. Use your creativity and make a flower that will "wow" the pollinator that you are trying to attract. You may use the back if you need more room. *(Example: A baseball bat and catcher's mitt flower to attract its' pollinator the baseball player. Do not use this example.)*

<u>DESIGN A FLOWER ACTIVITY TEACHER'S RUBRIC</u>

Criteria	1	2	3	4
Flower Parts	1 part	2 parts	3 parts	4 parts
Label and Description	Missing most labels and descriptions	Some labels and some descriptions are correct	All labels and most descriptions are correct.	All labels and descriptions are correct.
Creativity	Information is taken directly from text or notes with no additional personal touch.	Information is taken mostly from text or notes with a little personal design or wording.	Information is presented mostly in your own design and wording with key terms from text or notes.	Information is presented in an exceptionally creative yet accurate way.

Points scored on rubric	Grade
12	100
11	95
10	90
9	85
8	80
7	75
6	70
5 or below	65

SIXTH GRADE LESSON PLAN

Radical Roots

SUBJECT / GRADE LEVEL: 6th Grade - Science.

ACADEMIC STANDARDS AND PERFORMANCE INDICATORS:

Science:

6.L.5B.3 Develop and use models to compare structural adaptations and processes that flowering plants use for defense, survival and reproduction.

LESSON OBJECTIVES:

TSW identify the differences between fibrous and taproots. TSW will measure the growth of a taproot over a four week period.

MATERIALS:

- *Radical Roots* teacher guide (attached)
- *Radical Roots* student worksheet (attached)
- *Radish Race* Data Collection sheet (attached)
- Radish seeds

ENGAGEMENT:

- Draw a growing carrot. Label the roots, leaves, and stems.

EXPLORATION:

- The teacher will begin by explaining that the root of the carrot is the part that we eat.

EXPLANATION:

- The teacher will follow up the engagement question with fibrous and taproot descriptions from the *Radical Roots* teacher guide. The teacher will guide students through the completion of the *Radical Roots* student worksheet.
- The teacher will show *Growing Radishes* by Burpee Gardens: https://www.youtube.com/watch?v=O--t2wiL6mYw.

- Students will each plant a radish seed. The seedlings should be measured and recorded weekly for thirty days. These observations and measurements will be recorded on the *Radish Race* Data Collection sheet. It can be fun to turn it into a contest to see who has the tallest and widest plant at the end of the four week period.

EVALUATION:

- At the end of the thirty day growth period radishes should be planted in the garden to increase beneficial fungi in the soil. See *A Fungus Among Us* lesson for more details.

ELABORATION:

- At the end of the thirty day growth period radishes should be planted in the garden to increase beneficial fungi in the soil. See *A Fungus Among Us* lesson plan for more details.

RESOURCES:

1. *Growing Radishes* by Burpee Gardens: https://www.youtube.com/watch?v=O–t2wiL6mYw.
2. *Radical Roots* teacher guide (attached).
3. *Radical Roots* student worksheet (attached).
4. *Radish Race Data Collection* sheet (attached).

TEACHER GUIDE
RADICAL ROOTS

BACKGROUND:

Students should understand that roots absorb water and dissolved nutrients, anchor the plant and store food.

ENGAGEMENT:

- Ask students to draw a growing carrot labeling the roots, leaves, and stems.

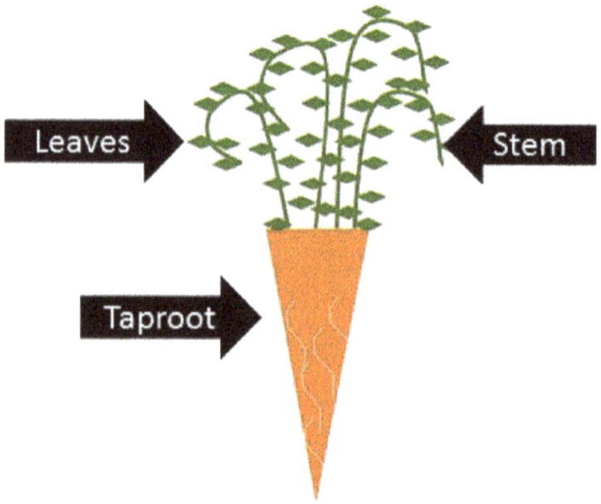

EXPLORATION:

Guide students through the completion of the *Radical Roots* student worksheet using the drawing above as well as the descriptions below.

Taproot

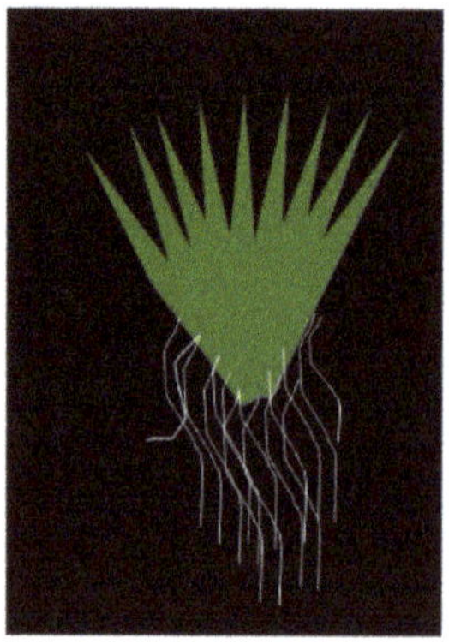

Fibrous Root

- *Fibrous Root*: has numerous uniform branching roots. Sometimes the fibrous root system can contain millions of roots. These root systems help to prevent erosion.
- *Taproot*: a primary root that grows vertically downward and gives off small lateral roots.

EXAMPLES OF FIBROUS AND TAPROOTS

- *Fibrous*: Grasses, Lily, Corn, Dandelion, Tomato, Lettuce.
- *Taproots*: Turnip, Beet, Carrot, Radish, Parsnip.

RADICAL ROOTS

Name ____________________________

Instructions: Based on the descriptions below, correctly label the taproot and the fibrous root.

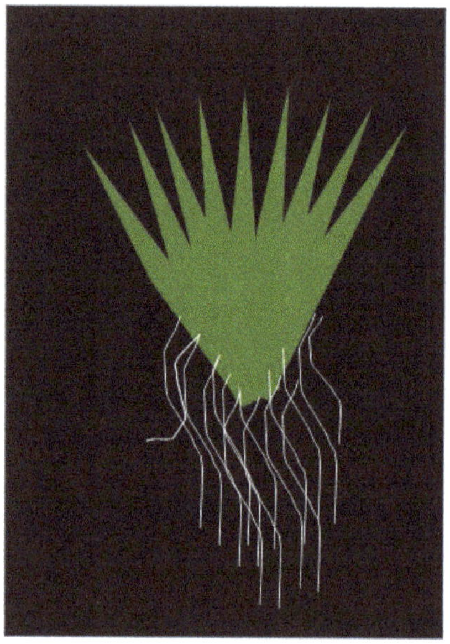

A) ____________________ B) ____________________

- **Fibrous Root** – has numerous uniform branching roots. Sometimes the fibrous root system can contain millions of roots. These root systems help to prevent erosion.

- **Taproot -** a primary root that grows vertically downward and gives off small lateral roots.

List three example of taproots:

1) ______________________________

2) ______________________________

3) ______________________________

List three example of fibrous roots:

1) ______________________________

2) ______________________________

3) ______________________________

RADISH RACE

Name ____________________________

Instructions: Observe, measure, and record the growth of your radish plant over the next four weeks.

	Week 1	Week 2	Week 3	Week 4
Height				
Width				
Positive Observation				
Negative Observation				

Instructions: Plot the growth in height of your radish plant over the next four weeks.

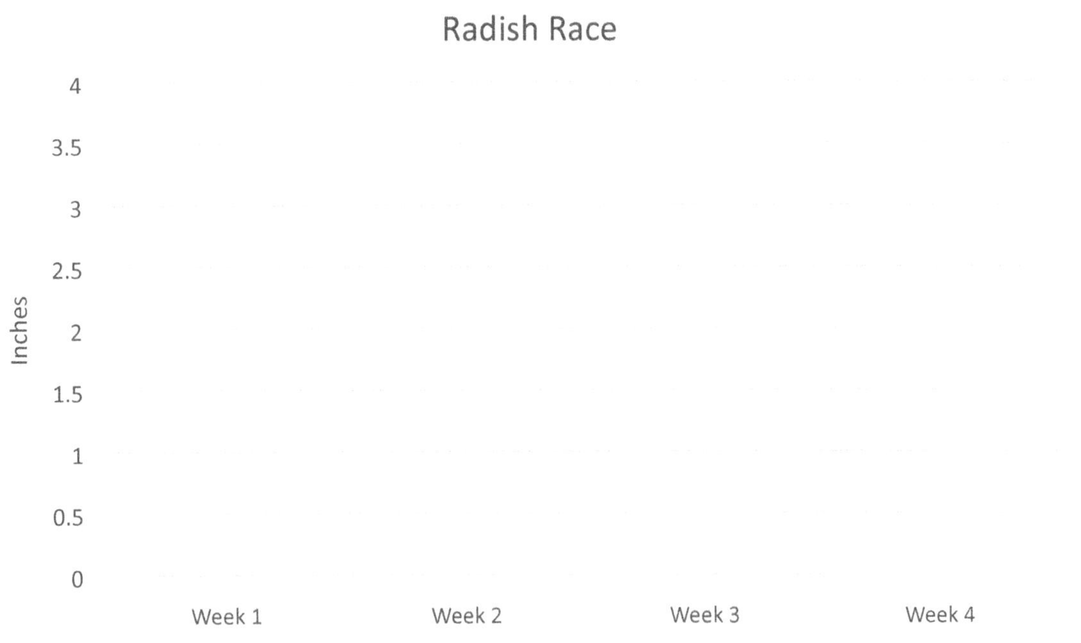

SIXTH GRADE LESSON PLAN
Spectacular Spores

SUBJECT / GRADE LEVEL: 6th Grade -Science.

ACADEMIC STANDARDS AND PERFORMANCE INDICATORS:

Science:

6.L.5 The student will demonstrate an understanding of the structures, processes, and responses that allow protists, fungi, and plants to survive and reproduce.

LESSON OBJECTIVES:

TSW be able to effectively distinguish between a seed and a spore. TSW identify the conditions necessary for spore germination.

MATERIALS:

- *Spectacular Spores* student worksheet (attached)
- *Spectacular Spores* teacher guide (attached)

ENGAGEMENT:

- **What is the fastest living organism on the planet?**

EXPLORATION:

- Teacher will follow up the engagement question with this video that answers the question: *Fungus Cannon In Super Slow Motion* by Earth Unplugged: https://www.youtube.com/watch?v=T8OAmcUnm4g.

EXPLANATION:

- The teacher will follow up the video with the spore and seed comparisons from the *Spectacular Spores* teacher guide. The teacher will guide students through the completion of the *Spectacular Spores* student worksheet.

- After completion of the student worksheet, teacher should show the following video: *Wow Fungi plant growth: The Private Life of Plants* by BBC Worldwide: https://www.youtube.com/watch?v=puDkLFcCZyI.
- The final part of this lesson is a virtual field trip to the mushroom production facility at *Mepkin Abbey: Mepkin Abbey Mushroom Production* by Steve Maxwell: https://www.youtube.com/watch?v=CaqlPglJueE.

EVALUATION:

- Students will turn in *Spectacular Spores* student worksheet.

ELABORATION:

- An extension to this lesson would be a mushroom growing kit for the classroom.

RESOURCES:

1. *Wow Fungi plant growth: The Private Life of Plants, from David Attenborough* by BBC: https://www.youtube.com/watch?v=puDkLFcCZyI.
2. *Fungus Cannon In Super Slow Motion* by Earth Unplugged: https://www.youtube.com/watch?v=T8OAmcUnm4g.
3. *Mepkin Abbey Mushroom Production* by Steve Maxwell: https://www.youtube.com/watch?v=CaqlPglJueE.
4. *Spectacular Spores* teacher guide (attached).
5. *Spectacular Spores* student worksheet (attached).

TEACHER GUIDE
SPECTACULAR SPORES

BACKGROUND:

Students should understand that plant reproduction began with the simple cell division of algae and has evolved to the complex reproductive structure of a seed. In between the algae and the seeds came spores. They function similarly to seeds but there are some important differences.

ENGAGEMENT:

- Teacher will ask students, "**What is the fastest living organism on the planet?**"

EXPLORATION:

Teacher will guide students through the completion of their *Spectacular Spores* worksheet using the Spores vs. Seeds information below:

SPORES VS. SEEDS FERTILIZATION:

- Spores: fertilization takes place after the spore leaves the parent plant. Self fertilization is more likely with spores than with pollination, which leads to seeds.
- Seeds: fertilization takes place before the seed leaves the parent plant. Because self pollination is less likely, seeds offer more opportunities for genetic diversity.

SIZE:

- Spores: require magnification to be seen.
- Seeds: are larger than spores and can easily be seen and touched.

CELLS:

- Spores: are unicellular.
- Seeds: are multicellular.

LOCATION:

- Spores: are located on the underside of non-flowering plants. Spores may be spread by wind but require a wet environment in order to germinate and grow.
- Seeds: are located either in the fruit or flower of flowering plants. Seeds germinate more easily than spores because they can germinate on dry land.

MUSHROOMS IN THE GARDEN:

- Mushrooms often appear in the garden after it rains because the rainwater provided the wet conditions that the spores needed to germinate. Mushrooms indicate that your garden has plenty of organic material in the soil. Mushrooms help break down that organic material, and make your soil more productive.

SPECTACULAR SPORES

Name_______________________

Part I: Why do mushrooms appear in the garden after it rains?

Part II: What do mushrooms in your garden indicate?

Part III: *Complete the chart below comparing and contrasting spores and seeds*:

	Spores	Seeds
Fertilization		
Size		
Cells		
Location		

SIXTH GRADE LESSON PLAN
Traveling Seeds

SUBJECT / GRADE LEVEL: 6th Grade - Science.

ACADEMIC STANDARDS AND PERFORMANCE INDICATORS:

Science:

6.L.5 The student will demonstrate an understanding of the structures, processes, and responses that allow protists, fungi, and plants to survive and reproduce.

6.L.5B.3 Develop and use models to compare structural adaptations processes that flowering plants use for defense, survival and reproduction.

LESSON OBJECTIVES:

TSW will compare the different types of seed adaptations that allow them to travel and germinate when and where conditions are best suited for growth.

MATERIALS:

- *Traveling Seeds Collection and Observation* worksheet (attached)
- *Traveling Seeds* teacher guide (attached)
- *Teaching Rubric* (attached)

ENGAGEMENT:

- The teacher will begin this lesson by posing the question, "**Since plants can't walk, how do they travel?**"

EXPLORATION:

- The teacher will begin by having students list the different strategies that plants use to disperse their seeds far from the parent plant fire, water, animals, wind, and explosions.

EXPLANATION:

- The teacher will follow up the engagement question with the engaging video titled *Seed Dispersal: The Great Escape* by Naturalist Outreach: https://www.youtube.com/watch?v=xY4JFOSuqvY.
- Using the *Traveling Seeds* teacher guide, the teacher will guide students through the completion of the *Traveling Seeds Collection and Observation* worksheet. Students will need to go outside to complete the worksheet.

ELABORATION:

- An extension to this lesson would be planting collected seeds from native plants to create a native wildlife habitat area.

EVALUATION:

- Students will turn in *Traveling Seeds Collection and Observation* worksheet.

RESOURCES:

1. *Seed Dispersal: The Great Escape* by Naturalist Outreach: https://www.youtube.com/watch?v=xY4JFOSuqvY.
2. *Traveling Seeds Collection and Observation* worksheet (attached).
3. *Traveling Seeds* teacher guide (attached).
4. *Traveling Seeds* teacher s rubric (attached).

TEACHER GUIDE
TRAVELING SEEDS

BACKGROUND:

Students should understand that seeds need to travel away from their parent plant or they would be left to compete with the parent for light, water, sun and nutrients.

ENGAGEMENT:

- Ask students to list three different strategies that plants use to disperse their seeds far from the parent plant.

EXPLORATION:

Guide students through the completion of their *Traveling Seeds* collection and observation worksheet using the Seed Dispersal Strategy information below:

- EXPLOSIONS: Explosions guarantee that the seeds will move far away from the parent plant. Many of these plants grow in full sun as they need the seedpods to dry out completely in order to disperse the seeds. Once the seedpod has completely dried out it then bursts open and the seeds shoot out in many directions.
- WIND: Some plants, such as red maple, have winged or "helicopter" seeds that are lifted into the air by the wind and flutter to the ground after traveling through the air. Other plants such as dandelion and salt myrtle have feathery seeds that can travel long distances by the wind.
- ANIMALS: Animals can spread seeds by eating them or sometimes seeds latch onto their skin, fur or feathers. This includes us as we can also spread seed that latches on to our clothing and shoes. We call these seeds "hitchhikers". Many of our favorite fruits are designed to be delicious so that animals will eat them and then later disperse the seeds through their droppings. Have you ever seen a plant growing somewhere odd and you wondered how the seed was planted in such a strange location? Chances are the seed was contained in bird droppings and it began to grow where it landed.
- WATER: Seeds that disperse through water are designed to float. They often have a thick seed coat that enables them to float long distances without breaking down. Many of the plants from these seeds are found growing along pond and river banks.
- FIRE: When a forest fire begins, plants have no way to escape except through an adaptation in which they have developed a way for their seeds to survive a fire. In fact, some conifers require the heat of a fire in order for their cones to open up and disperse the seeds from within.

Teaching Rubric for Traveling Seeds Lesson

Criteria	1	2	3	4
Seeds Collected	4 or less seeds	5 seeds	6 seeds	7 seeds
Description	All Seed descriptions are missing	Most seed descriptions are missing or lacking detail	Some seed descriptions are detailed	All seeds descriptions are detailed
Sketch	All sketches are missing	Most sketches are missing	Some sketches are presented in an exceptionally creative yet accurate way.	All sketches are presented in an exceptionally creative yet accurate way.

Points scored on rubric	Grade
12	100
11	95
10	90
9	85
8	80
7	75
6	70
5 or below	65

TRAVELING SEEDS Name ______________________________

Instructions: List five strategies that plants use to disperse their seeds far from the parent plant.

1) ________________________ **2)** ________________________ **3)** ________________________

4) ________________________________ **5)** ________________________________

Application: Collect seven different seeds types. Complete the table below based on the seeds that you collected.

	Dispersal Strategy (Fire, water, animals, wind, explosion)	Description (Round, winged, thick seed coat, etc.)	Sketch (Sketch the shape of the seed- if possible use color)
Seed 1			
Seed 2			
Seed 3			
Seed 4			
Seed 5			
Seed 6			
Seed 7			

SIXTH GRADE LESSON PLAN
Trusty Tropisms

SUBJECT / GRADE LEVEL: 6th Grade - Science - Duration: Two sixty minute lessons.

ACADEMIC STANDARDS AND PERFORMANCE INDICATORS:

Science:

6.L.5 The student will demonstrate an understanding of the structures, processes, and responses that allow protists, fungi, and plants to survive and reproduce.

LESSON OBJECTIVES:

TSW differentiate between the different tropic responses to stimuli that plants have adapted to survive.

MATERIALS:

- *Trusty Tropisms* student worksheet and teacher guide (attached)
- *Build A Plant Maze Experiment and Observation* sheet (attached)
- Cardboard, shoeboxes, tape, and sugar snap pea seeds

ENGAGEMENT:

- **How can vines climb without arms or legs?**

EXPLORATION:

- The teacher will explain that vines climb through a response called *thigmotropism*.

EXPLANATION:

- **DAY 1**: The teacher will follow up the engagement question with the video: *Biology Lesson Idea: Plants, Tropisms and Hormones* by Twig World: https://www.youtube.com/watch?v=pCFstSMvAMI.

- Using the *Trusty Tropisms* teacher guide (attached) , the teacher will guide students through the completion of the *Trusty Tropisms* student worksheet (attached). Teacher will instruct students to bring in supplies for the *Build A Plant Maze Experiment* (attached).
- **DAY 2**: Students will construct *Plant Maze Experiment* using the *Build A Plant Maze Experiment and Observation* sheet (attached).

ELABORATION:

- At the end of the *Build A Plant Maze* experiment, peas can be planted directly into the garden. These legumes will add nitrogen to your soil as they grow.

EVALUATION:

- **DAY 1**: Students will turn in *Trusty Tropisms* student worksheet.
- **DAY 2**: Students will record observations on their *Build A Plant Maze Experiment and Observation* sheet until their pea has reached the top of the maze (approximately 2-3 weeks).

RESOURCES:

1. *Biology Lesson Idea: Plants, Tropisms and Hormones* by Twig World: https://www.youtube.com/watch?v=pCFstSMvAMI.
2. *Trusty Tropisms* teacher guide (attached).
3. *Trusty Tropisms* student worksheet (attached).
4. *Build A Plant Maze Experiment* worksheet (attached).

TEACHER GUIDE
TRUSTY TROPISMS

BACKGROUND:

Students should understand that plants are not able to move to a new location once they begin growing where conditions are not favorable for plant growth. Therefore, in order to survive, plants must alter their growth in order to make conditions more favorable for plant growth.

ENGAGEMENT:

- Teacher will pose the question, "**How can vines climb without arms or legs?**"

EXPLORATION:

Guide students through the completion of using the drawing above as well as the descriptions below:

- A *tropic response* happens when a plant responds to stimuli in order to grow in a direction that will locate it a place more favorable for plant growth.
- The word *tropism* comes from the Greek word "trope" which means to change or turn. Plants are not able to move to a new location once they begin growing where conditions are not favorable. Therefore, in order to survive plants must alter their growth in order to make conditions more favorable for plant growth.
- *Phototropism* happens when a plant stem grows towards the light. A plant hormone called *auxin* is responsible for phototropism.
- *Thigmotropism* happens when a plants responds to a touch typically from a hard object such as a trellis or stake. This typically occurs in vines. The cells in the side of the plant that touches the hard object produce a plant hormone called *auxin*. The *auxin* then gets transferred to the side of the vine that is not touching the trellis or stake. This causes the untouched cells to grow faster and stretch out. When this happens, the vine begins to grow around the object.
- *Gravitropism* or *Geotropism* happens when a plant responds to gravity. *Gravitropism* causes roots to grow down and stems to grow up.

Plant Responses to Stimuli

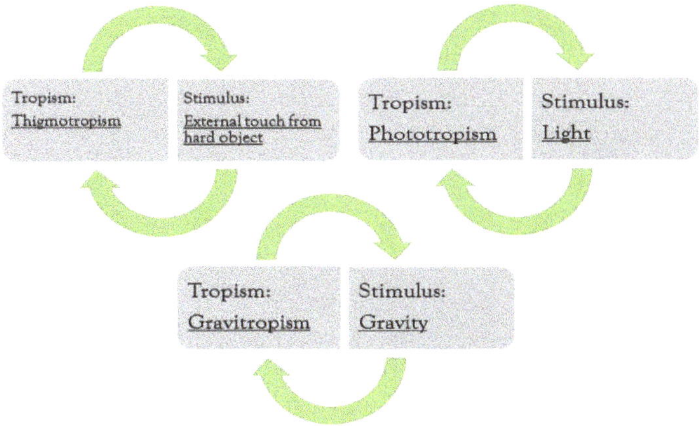

TRUSTY TROPISMS

Name ______________________

Instructions: Answer the questions below then use your answers to complete the graphic organizer.

1) What is a tropic response?

2) Why do we call these responses tropisms?

3) Why do plants have tropic responses?

4) What is phototropism?

5) What causes phototropism?

6) How does thigmotropism work?

7) What is gravitropism?

Plant Responses to Stimuli

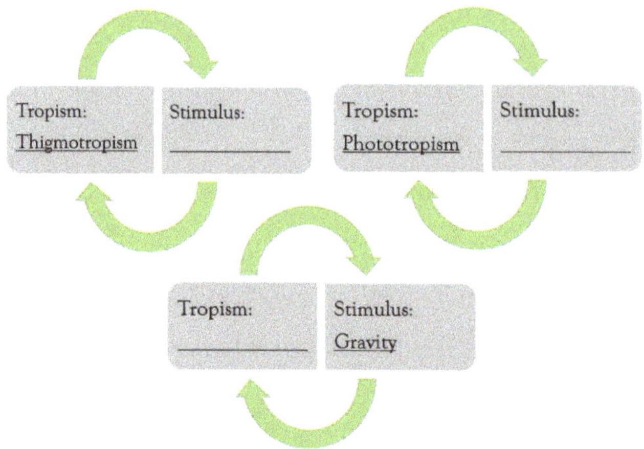

BUILD A PLANT MAZE

Name ____________________________

Instructions: Build a plant maze following the directions below. Use the chart at the bottom of the page to record your weekly observations and measurements as you watch your plant move through the maze exhibiting phototropism.

1) Stand your shoebox on end and cut a small 2" square hole at the top of the box.
2) Cut pieces of cardboard and attach them to the inside of the box so they stick out creating alternating partitions as you go from the bottom to the top of the box.
3) Plant pea seeds in a small container. Water them and then place them in the bottom of the shoebox.
4) Place the lid on the shoebox ensuring that it blocks out all light except for the light that comes through the hole in the top. Every few days remove the lid to give the growing plant some water then immediately put the lid back on.
5) Record and observe your plant weekly until it makes it out of the top of the maze thanks to phototropism.

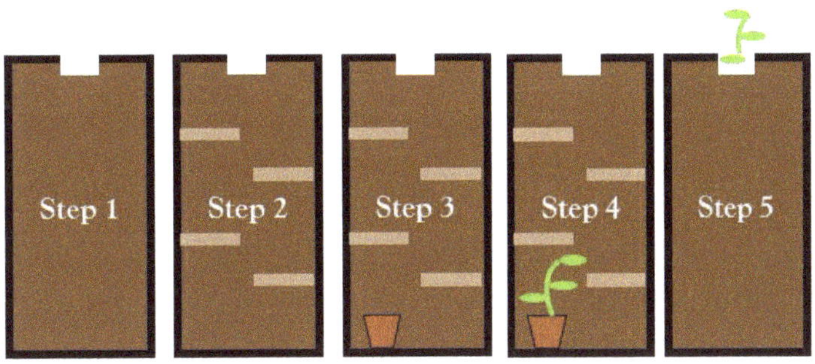

Instructions: Observe, measure, and record the growth of your pea plant over the next three weeks.

	Week 1	Week 2	Week 3	Final Observations and Measurements
Height of Pea Plant				
Distance from top of maze				
Positive Observation				
Negative Observation				

SIXTH GRADE LESSON PLAN
Amazing Adaptations

SUBJECT / GRADE LEVEL: 6th Grade -Science - Duration: Two sixty minute lessons.

ACADEMIC STANDARDS AND PERFORMANCE INDICATORS:

Science:

6.L.4 The student will demonstrate an understanding of how scientists classify organisms and how the structures, processes, behaviors, and adaptations of animals allow them to survive.

LESSON OBJECTIVES:

TSW compare the plant adaptations that have developed to allow plants to protect themselves as well as survive extreme conditions.

MATERIALS:

- *Amazing Adaptations* teacher guide (attached)
- *Amazing Adaptations* student worksheet (attached)
- *Ultimate Veggie Design Project and Rubric* (attached)

ENGAGEMENT:

- **How would you adapt to survive if your feet were glued to the ground?**

EXPLORATION:

- The teacher will begin by explaining that plants have developed adaptations to protect themselves as well as survive extreme conditions.

EXPLANATION:

- **DAY 1**: The teacher will follow up the engagement question with adaptation descriptions from the *Amazing Adaptations* teacher guide. The teacher will guide students through the completion of the *Amazing Adaptations* table on the student worksheet. The teacher will show *You At The Zoo – Plant Adaptations* by ThinkTVPBS: https://www.youtube.com/watch?v=fVfV5LMXh2s.

- DAY TWO: Students will complete *Ultimate Veggie Design Project.*

ELABORATION:

- Students can grow sensitive plants from seed and observe their reaction to human touch.

EVALUATION:

- **DAY 1**: Students will turn in the *Amazing Adaptations* student worksheet.
- **DAY 2**: Students will complete *Ultimate Veggie Design Project.*

RESOURCES:

1. *You At the Zoo – Plant Adaptations* by ThinkTVPBS: https://www.youtube.com/watch?v=fVfV5LMXh2s.
2. *Amazing Adaptations* teacher guide (attached).
3. *Amazing Adaptations* student worksheet (attached).
4. *Ultimate Veggie Design Project and Rubric* (attached).

TEACHER GUIDE
AMAZING ADAPTATIONS

BACKGROUND:

Students should understand that plants have developed adaptations to protect themselves as well as survive extreme conditions.

ENGAGEMENT:

- Teacher will pose this question to students, "**How would you adapt to survive if your feet were glued to the ground?**"

EXPLORATION:

Teacher will guide students through the completion of the *Amazing Adaptations* student worksheet using the adaptations descriptions below:

- DON'T TOUCH ME ADAPTATION: The sensitive plant closes when touched in order to avoid being eaten by animals. It can also close the leaves at night to conserve water.
- CHEMICAL ADAPTATION: Certain plants such as black walnut trees have a chemical in their leaves and roots. It is toxic when released into the soil. This prevents seedlings of other plant species from being able to grow near them thus ensuring that they keep all of the light, water and nutrients for themselves.
- WARNING SIGNAL: When the leaves of certain plants such as tomatoes are being eaten, the plants release a chemical to warn other plants of the same species nearby. The plants release a wounding hormone that not only heals the existing wounds but helps to prevent future wounding as well.
- CALLING FOR BACKUP: When the leaves of beans, corn, cucumbers and tomatoes are being eaten, they release chemicals which attract predators of the pests that are eating them.
- EPIPHYTES: Epiphytes grow above the ground and use other plants or objects for support. Their roots are in the air and they do not feed off of the plants on which they grow. However, by growing on other plants they manage to reach positions where they can get optimal light. Their roots absorb water and nutrients directly from the air.
- ANTIFREEZE: Certain conifers that live in very cold conditions actually have something in their sap that acts as an antifreeze. Therefore, no matter how cold it gets the plant can still survive the harsh conditions This antifreeze sap gives them their distinct pine scent. The waxy cuticle on the outside of the needles helps to prevent water loss as well as ensures that snow does not accumulate on the branches causing them to break.

AMAZING ADAPTATIONS Name________________________

Instructions: Complete the chart below with six different plant adaptations.

Plant	Adaptation Description

ULTIMATE VEGGIE DESIGN TEACHER'S RUBRIC

Criteria	1	2	3	4
Adaptation	0 adaptations	1 adaptation	2 adaptations	3 adaptations
Adaptation Label & Description	Missing all labels and descriptions	Missing most labels and descriptions	Adaptation labels and descriptions are present	Adaptation labels and descriptions are detailed
Creativity	Design lacks color and details	Design is adequate and colorful	Design is nicely drawn but lacks creativity and a unique personal touch	Design is colorful, unique, creative and shows a personal touch

Points scored on rubric	Grade
12	100
11	95
10	90
9	85
8	80
7	75
6	70
5 or below	65

ULTIMATE VEGGIE DESIGN Name________________________

Instructions: Design a vegetable for your garden that has three adaptations that will enable it to defend itself and survive. Label each adaptation appropriately and write at least one sentence describing what each adaptation does on your ultimate vegetable. Use your creativity and make a vegetable that that will not only survive but thrive! You may use the back if you need more room.

SIXTH GRADE LESSON PLAN
Capillary Celery

SUBJECT / GRADE LEVEL: 6th Grade - Science.

ACADEMIC STANDARDS AND PERFORMANCE INDICATORS:

Science:

6.L.5 The student will demonstrate an understanding of the structures, processes, and responses that allow protists, fungi, and plants to survive and reproduce.

6.L.5B.1 Construct explanations of how the internal structures of vascular and non vascular plants transport food and water.

LESSON OBJECTIVES:

TSW be able to explain the purpose of the xylem and phloem in a plant stem. TSW observe the capillary action that moves water and dissolved nutrients throughout a plant.

MATERIALS:

- *Capillary Celery* student worksheet (attached)
- *Capillary Celery* teacher guide (attached)
- Celery
- Food Coloring
- Knife
- Jar
- Water

ENGAGEMENT:

- **What do you do with the base of a celery plant once you've chopped of the stems to eat?**

EXPLORATION:

- The teacher will chop off the base of a celery plant and show students that it can be planted in a pot or directly into the garden. Students will observe celery base and it grow new leaves and stems over the following weeks.

EXPLANATION:

- The teacher will explain that the remaining celery stems will be used to demonstrate the movement of water through the xylem of the plant by means of capillary action. Teacher will then guide students through the completion of their *Capillary Celery* worksheet Part I.
- After completion of Part I, teacher will show the following video: *Plant Transport: Xylem and Phloem, Transpiration [3D Animation]* by Biology/Medicine Animations HD: https://www.youtube.com/watch?v=xGCnuXxbZGk.
- Teacher will place celery stems in a jar with red or blue colored water. Students will complete *Capillary Celery* worksheet Part II.

ELABORATION:

- An extension to this lesson would be to plant multiple bases of celery plants from the store in the garden and monitor their re-growth.

EVALUATION:

- Students will turn in *Capillary Celery* student worksheet.

RESOURCES:

1. *Plant Transport: Xylem and Phloem, Transpirations [3D Animation]* by Biology/Medicine Animations HD: https://www.youtube.com/watch?v=xGCnuXxbZGk.
2. *Capillary Celery* teacher guide (attached).
3. *Capillary Celery* student worksheet (attached).

TEACHER GUIDE
CAPILLARY CELERY

BACKGROUND:

Students should understand that vascular plants have conductive tissues that move food and water throughout the plant.

ENGAGEMENT:

- Teacher will ask students, **"What do you do with the base of a celery plant once you've chopped of the stems to eat?"**

EXPLORATION:

- Teacher will chop off the base of a celery plant and show students that it can be planted in a pot or directly into the garden.
- Students will observe celery base and it grow new leaves and stems over the following weeks.

EXPLANATION:

- Teacher will explain that the remaining celery stems will be used to demonstrate the movement of water through the *xylem* of the plant by means of *capillary action*.
- Teacher will then guide students through the completion of their *Capillary Celery* worksheet Part I using the following information:
- *Capillary Action*: the ability of a liquid to flow in narrow spaces without the assistance gravity.
- *Xylem*: moves water and dissolved minerals. The xylem dies after one year and develops new tissue the following years. The dead xylem accounts for the rings in tree trunks.
- *Phloem*: moves food in the form of sugars and other molecules created by the plant. Phloem is always alive.

	XYLEM	PHLOEM
DIRECTION	Moves upward only	Moves up and down the stem
LOCATION	Located at center of vascular bundle	Located at outside of vascular bundle
CELL WALLS	Thick and hard lignin cell walls	Thin cellulose walls
PERMEABILITY	Impermeable	Permeable

- Teacher will place celery stems in a jar with red or blue colored water.
- Students will complete *Capillary Celery* worksheet Part II.

EVALUATION:

- Students will turn in the *Capillary Celery* worksheet.

CAPILLARY CELERY

Name ____________________________

Part I: *Answer the following questions as well as complete the chart below.*

1) What is capillary action?

____

____

2) What is xylem?

____

____

3) What is phloem?

____

____

	Xylem	**Phloem**
Direction		
Location		
Cell Walls		
Permeability		

Part II: *Place celery stems in a jar with red or blue colored water. Predict the amount of time it will take for the colored water to move upward through the xylem due to capillary action.*

Prediction

How tall (*inches*) is the stalk of celery in your jar?

____

How much time do you predict it will take for the xylem to move the water one inch up the stem?

____

Based on your two previous answers, how long do you predict it will take for the xylem to move the water up the entire stem?

____

Conclusion

How long did it actually take for the xylem to move the water up the entire stem?______________________

SIXTH GRADE LESSON PLAN
A Fungus Among Us

SUBJECT / GRADE LEVEL: 6th Grade - Science.

ACADEMIC STANDARDS AND PERFORMANCE INDICATORS:

Science:

6.L.5 The student will demonstrate an understanding of the structures, processes, and responses that allow protists, fungi, and plants to survive and reproduce.

LESSON OBJECTIVES:

TSW be able to describe both the harm and benefits of fungus in the garden. TSW create a plan to increase beneficial fungi in the garden.

MATERIALS:

- *A Fungus Among Us* student worksheet (attached)
- Radish plants (optional)

ENGAGEMENT:

- Students will describe what might happen if a garden was infected with a fungus.

EXPLORATION:

- The teacher will instruct students to complete Part I and II of the worksheet indicating fungus to be either harmful or beneficial in the garden.

EXPLANATION:

- The teacher will introduce Part II with the video titled: *Beneficial Fungi* by Oklahoma Gardening: https://www.youtube.com/watch?v=-BQEFl5RiVc.
- After showing beneficial fungi video, the teacher will show video explaining the spread of harmful plant fungi entitled *Fungicide No. 2: How Do Fungi Attack Plants?* by Purdue Agriculture: https://www.youtube.com/watch?v=eGwbUzPX4jE.

ELABORATION:

- An extension to this lesson would be planting radishes to increase beneficial fungi in the soil. This lesson can be used as a follow-up to a *Race to Radishes* lesson. Plants from the *Race to Radishes* lesson can be planted in the ground as an extension to this lesson.

EVALUATION:

- Students will turn in *A Fungus Among Us* worksheet.

RESOURCES:

1. *Beneficial Fungi* by Oklahoma Gardening: https://www.youtube.com/watch?v=-BQEFl5RiVc.
2. *Fungicide No. 2: How Do Fungi Attack Plants?* by Purdue Agriculture: https://www.youtube.com/watch?v=eGwbUzPX4jE.
3. *A Fungus Among Us* student worksheet (attached).

A FUNGUS AMONG US

Name ______________________

Part I: Circle the word that best describes fungus in the garden:

HARMFUL or BENEFICIAL

Part II: *After watching both videos, explain how fungus can be harmful and beneficial in the garden.*

____

____

____

Part III: *How can you increase beneficial Mycorrhizal fungi in your garden?*

____

____

Mycorrhizal Fungi

This amazing fungus comes from the Greek 'mykos' meaning fungus and 'riz' meaning roots. These naturally occurring fungi can help your plants absorb nutrients as well as survive drought. Over ninety percent of plants have a fungus associated with their roots and most rely on their fungal partner for survival. You can increase beneficial fungi in your garden by planting a cover crop such as radishes.

Investigate: How many radishes can you grow in one square foot?

____

Measure: How many square feet do you have in your garden? (*Length x Width*)

____

Calculate: How many radishes will fit in your garden?

____

Plan: In the area below, sketch your garden and indicate how much space you would be able to devote to growing radishes in order to increase beneficial fungi in your soil.

SIXTH GRADE LESSON PLAN
Pesky Pests

SUBJECT / GRADE LEVEL: 6th Grade - Science.

ACADEMIC STANDARDS AND PERFORMANCE INDICATORS:

Science:

6.L.4 The student will demonstrate an understanding of how scientists classify organisms and how the structures, processes, behaviors, and adaptations of animals allow them to survive.

LESSON OBJECTIVES:

TSW identify different types of garden pests. TSW evaluate the current pest population in their school garden.

MATERIALS:

- *Pesky Pests* student worksheet (attached)
- *Pesky Pests* teacher guide (attached)
- *Pest Scouting Observation* Sheet (attached)

ENGAGEMENT:

- **Where do pests exist in nature?**

EXPLORATION:

- The teacher will explain that pests do not exist in nature. Merriam Webster defines a *pest* as a plant or animal detrimental to humans or human concerns. Pests do not exist in nature. Pests exist merely because they cause detriment to humans.

EXPLANATION:

- Using the *Pesky Pests* Teacher Guide, the teacher will guide students through the completion of the *Pesky Pests* student worksheet.
- The teacher will follow up the *Pesky Pests* student worksheet with the video *Parasitic Wasps* by National Geographic: https://www.youtube.com/watch?v=Bc69LLLEQRk.

- After completing the worksheet, students will go to the garden in order to complete the *Pest Scouting Observation* sheet.

ELABORATION:

- Students can set up yellow sticky traps to monitor and control insects in the garden.

EVALUATION:

- Students will turn in *Pesky Pests* student worksheet.
- Students will record observations on their *Pest Scouting Observation* sheet.

RESOURCES:

1. Dictionary Definition: http://www.merriam-webster.com/dictionary/pest.
2. *Parasitic Wasps* by National Geographic: https://www.youtube.com/watch?v=Bc69LLLEQRk.
3. *Pesky Pests* teacher guide (attached).
4. *Pesky Pets* student worksheet (attached).
5. *Pest Scouting Observation* sheet (attached).

TEACHER GUIDE
PESKY PESTS

BACKGROUND:

Students should understand that a pest does not have to be an insect. It can be an animal, a disease, weeds or even humans.

ENGAGEMENT:

- Teacher will pose the question, "**Where do pests exist in nature?**"

EXPLORATION:

Guide students through the completion of the *Pesky Pests* student worksheet using the information below:

- **What is a pest?** Pests do not exist in nature. Merriam Webster defines a pest as a plant or animal detrimental to humans or human concerns. Pests do not exist in nature. Pests exist merely because they cause detriment to humans.
- **What is a pesticide?** A pesticide is any substance used to kill, repel, or control living organisms that are considered to be pests.

PESTICIDE	PESTS CONTROLLED
Insecticide	Insects
Fungicide	Fungi
Herbicide	Weeds
Rodenticides	Rodents
Miticides	Mites
Molluscicides	Slugs and Snails

- **What is a biological control?** Biological controls reduce pest populations by using natural enemies often introduced into the garden by humans.

BIOLOGICAL CONTROL	PESTS CONTROLLED
Parasitic Wasp	Aphids
Beneficial Nematodes	Grubs, Fungus Gnats, Fleas
Ladybug	Aphids, Chinch Bugs, Thrips
Praying Mantis	Aphids, Leafhoppers, Mosquitoes,
Caterpillar	Green Lacewing Aphids, Mites, Thrips, Whitefly

- *Scouting*: The process of assessing the pest population in the garden to evaluate in order to determine the necessary pest control measures if any.

Pesky Pests Name ___________________________

Instructions: Complete the tables and questions below.

What is a pest? ____
____
____

Pest	Is a pest when:	Is not a pest when:
Racoon	It gets into your garbage	Living in the woods

What is a pesticide? ____
____
____

Pesticide	Pests Controlled

What is a biological control? ____
____
____

Biological Control	Pests Controlled

PEST SCOUTING

Name ___________________________

Instructions: You have been hired as a pest scout to assess the pest population in your garden. Use the table below to record the pest activity in your garden. Just because you do not see the pests does not mean that they are not there. Make sure you record evidence of pest activity as well as actual pests. See the lettuce example below to get started.

Crop	Weeds	Diseases	Insect
Lettuce	No Weeds Present	Yellow spots on leaves/possible disease	No insects present but holes in leaves indicate insect damage

SIXTH GRADE LESSON PLAN
Scavenger Hunt

SUBJECT / GRADE LEVEL: 6th Grade - Science.

ACADEMIC STANDARDS AND PERFORMANCE INDICATORS:

Science:

6.L.4 The student will demonstrate an understanding of how scientists classify organisms and how the structures, processes, behaviors, and adaptations of animals allow them to survive.

6.L.5 The student will demonstrate an understanding of the structures, processes, and responses that allow protists, fungi, and plants to survive and reproduce.

LESSON OBJECTIVES:

TSW identify different types of garden pests. TSW identify the female and male reproductive parts of a flowering plant. TSW identify the differences between fibrous and taproots. TSW be able to effectively distinguish between a seed and a spore. TSW will compare the different types of seed adaptations that allow them to travel and germinate when and where conditions are best suited for growth. TSW differentiate between the different tropic responses to stimuli that plants have adapted to survive.

MATERIALS:

From Previous Lessons:

- *Pesky Pests* student worksheet
- *Radical Roots* student worksheet
- *Spectacular Spores* student worksheet
- *Traveling Seeds* student worksheet
- *Powerful Pollinators* student worksheet
- *Trusty Tropisms* student worksheet
- *Scavenger Hunt* student worksheet (attached)
- Paper lunch sacks to collect items

*****Students should complete the lessons listed above to gain prior knowledge in order to complete the scavenger hunt.***

ENGAGEMENT:

- **How many plant terms can you name?**

EXPLORATION:

- The teacher will poll students for plant terms until all of the terms for the scavenger hunt have been stated.

EXPLANATION:

- Using the *Scavenger Hunt* chart, the teacher will guide students through the explanation of all terms in the chart. After reviewing the *Scavenger Hunt* chart, students will go to the garden in order to find the items.

ELABORATION:

- Students can compost all of the items from the scavenger hunt including the chart and the paper bag. If there is nowhere to compost the items, this could also be turned into a digital scavenger hunt where students simply take pictures of the items in the chart.

EVALUATION:

- Students will show teacher items in the bag as the teacher checks them off of the *Scavenger Hunt* chart.

RESOURCES:

1. Previous Lesson Plans.
2. *Scavenger Hunt* student worksheet (attached).

SCAVENGER HUNT

Name ___________________________

Instructions: Find all of the items listed below:

Item	Where did you find it?
1) Sepals	
2) A seed designed to spread by wind	
3) A plant that exhibits thigomotropism	
4) Taproot	
5) Fibrous Root	
6) Anther and/or Filament	
7) Stigma and/or Style	
8) Petals	
9) Something that reproduces with spores	
10) Seed that latches to animals	

LESSONS FOR SEVENTH GRADE

SEVENTH GRADE LESSON PLAN
Companion Plantings

SUBJECT / GRADE LEVEL: 7th Grade - Science.

ACADEMIC STANDARDS AND PERFORMANCE INDICATORS:

Science:

7.EC.5 The student will demonstrate an understanding of how organisms interact with and respond to the biotic and abiotic components of their environments.

7.EC.5B.1 Develop and use models to explain how organisms interact in a competitive or mutually beneficial relationship for food, shelter, or space including competition, mutualism, commensalism, parasitism, and predator-prey relationships.

LESSON OBJECTIVES:

TSW discover the complex interactions between organisms in the environment. TSW examine the benefits of mutualism in the garden provided by companion plantings.

MATERIALS:

- *Companion Plantings* student worksheet and *Planting Chart* (attached)
- *Companion Plantings* teacher guide (attached)

ENGAGEMENT:

- **What is mutualism?**

EXPLORATION:

- The teacher will begin by **asking students to envision a growing carrot with a taproot that goes deep into the soil.** The thin, feather-like leaves soak up lots of sunshine. Now you need to figure out what to grow next to the carrots. If it also has deep roots, than it will compete with the carrots for water and nutrients below the soil. If its leaves are taller, they will shade the carrots from getting the sun they need. **What characteristics would be good in a plant that would live next to the carrots?**

Answer: A plant with shallow roots, that also stays short so it won't block the sunlight from the carrots. Lettuce would be a good option.

EXPLANATION:

- DAY ONE: The teacher will follow up the engagement and exploration activities with the video titled *Companion Planting* by GrowOrganic Peaceful Valley: https://www.youtube.com/watch?v=G_Acdv-IkdI.
- Using the *Companion Plantings* teacher guide, the teacher will guide students through the completion the *Companion Plantings* student worksheet and chart.
- The teacher will encourage students to explore additional companion plantings for homework to be shared with the class on the following day.

ELABORATION:

- If the teacher would like to extend the lesson, the class can plan and plant companion plantings in the school garden.

EVALUATION:

- Students will turn in the *Companion Plantings* student worksheet and chart.

RESOURCES:

1. *Companion Planting* by GrowOrganic Peaceful Valley: https://www.youtube.com/watch?v=G_Acdv-IkdI.
2. General Resource: https://www.garden.org/ediblelandscaping/?page=201005-interplanting.
3. *Companion Plantings* teacher guide (attached).
4. *Companion Planting* student worksheet and *Planting Chart* (attached).

TEACHER GUIDE
COMPANION PLANTINGS

BACKGROUND:

Students should understand that even though plants may seem crowded together in a forest, the forest remains healthy because the plants there have mutualistic relationships and like to grow together. We can mimic this in the garden by placing plants that can work together and help each other.

ENGAGEMENT:

- Teacher will pose the question, "**What is mutualism?**"

EXPLANATION:

Guide students through the completion of the *Companion Plantings* student worksheet using the descriptions below.

- *Mutualism*: The way two organisms of different species exist in a relationship in which each individual benefits from the activity of the other.
- *Companion Planting*: The idea that plants can help in other when you purposefully plant them close to each other.

BENEFITS OF COMPANION PLANTINGS:

- Companion plantings can attract beneficial insects that can act as predators of harmful pests.
- Companion plants, such as onions, can repel pests away.
- Companion plantings make maximum use of garden space by allowing taller plants to be used as physical support for vining plants.
- Companion plantings can provide shelter from wind or too much shade.
- Companion plants can improve the soil conditions for other plants in the garden.

EVALUATION:

- Teacher will direct students to the following website in order to create their *Companion Planting* chart: http://garden.org/ediblelandscaping/?page=201005-interplanting.

Companion Plantings

Name ___________________________

Part I: Complete the following questions:

1) What is mutualism?

____

____

2) What is companion planting?

____

____

3) List five benefits of companion planting:

   a) ____

   b) ____

   c) ____

   d) ____

   e) ____

Part II: Create a companion planting chart based on the information provided to you in this lesson.

PLANT	COMPANION	INCOMPATIBLE

SEVENTH GRADE LESSON PLAN
Edible Ecosystem

SUBJECT / GRADE LEVEL: 7th Grade - Science - Duration: Two sixty minute lessons.

ACADEMIC STANDARDS AND PERFORMANCE INDICATORS:

Science:

7.EC.5 The student will demonstrate an understanding of how organisms interact with and respond to the biotic and abiotic components of their environments.

7.EC.5A.1 Develop and use models to describe the characteristics of the levels of organization within ecosystems including species, populations, communities, ecosystems, and biomes.

LESSON OBJECTIVES:

TSW discover the array of combinations of plants that can be put together in naturalistic patterns that create mutually beneficial relationships creating a garden ecosystem.

MATERIALS:

- *Edible Ecosystem* student worksheet (attached)
- *Edible Ecosystem* teacher guide (attached)
- *Design An Edible Ecosystem Project* (attached)

ENGAGEMENT:

- **What is an ecosystem?**

EXPLORATION:

- The teacher will begin by writing the words stinky, smelly and gross to describe the fragrance of a flower.

EXPLANATION:

- DAY ONE: The teacher will follow up the engagement question with the engaging video titled *Food Forests Natural Building with Permaculture Jax* by Sustainable North Florida: https://www.youtube.com/watch?v=0ywhC3pdAJ0.
- Using the *Edible Ecosystem* teacher guide, the teacher will guide students through the completion of the *Edible Ecosystem* student worksheet Part I and II.
- DAY TWO: Students will work independently on *Design An Edible Ecosystem Project.*

EVALUATION:

- DAY ONE: Students will turn in the *Edible Ecosystem* student worksheet.
- DAY TWO: Students will turn in *Design An Edible Ecosystem* project.

ELABORATION:

- If the teacher wished to extend the lesson, the class can create an edible ecosystem at the school.

RESOURCES:

1. *Food Forests Natural Building with Permaculture Jax* by Sustainable North Florida: https://www.youtube.com/watch?v=0ywhC3pdAJ0.
2. *Edible Ecosystem* teacher guide (attached).
3. *Edible Ecosystem* student worksheet (attached).
4. *Design an Edible Ecosystem Project* and *Rubric* (attached).

TEACHER GUIDE
EDIBLE ECOSYSTEM

BACKGROUND:

Students should understand that edible ecosystems are plants that are put together in naturalistic patterns that create mutually beneficial relationships creating a garden ecosystem.

ENGAGEMENT:

- Teacher will pose question, "**What is an ecosystem?**"

EXPLANATION:

Guide students through the completion of the *Edible Ecosystem* student worksheet using the drawing and descriptions below:

BENEFITS OF A FOOD FOREST EDIBLE ECOSYSTEM:

- Mostly self–maintaining.
- High yields of many foods.
- A healthy ecosystem.

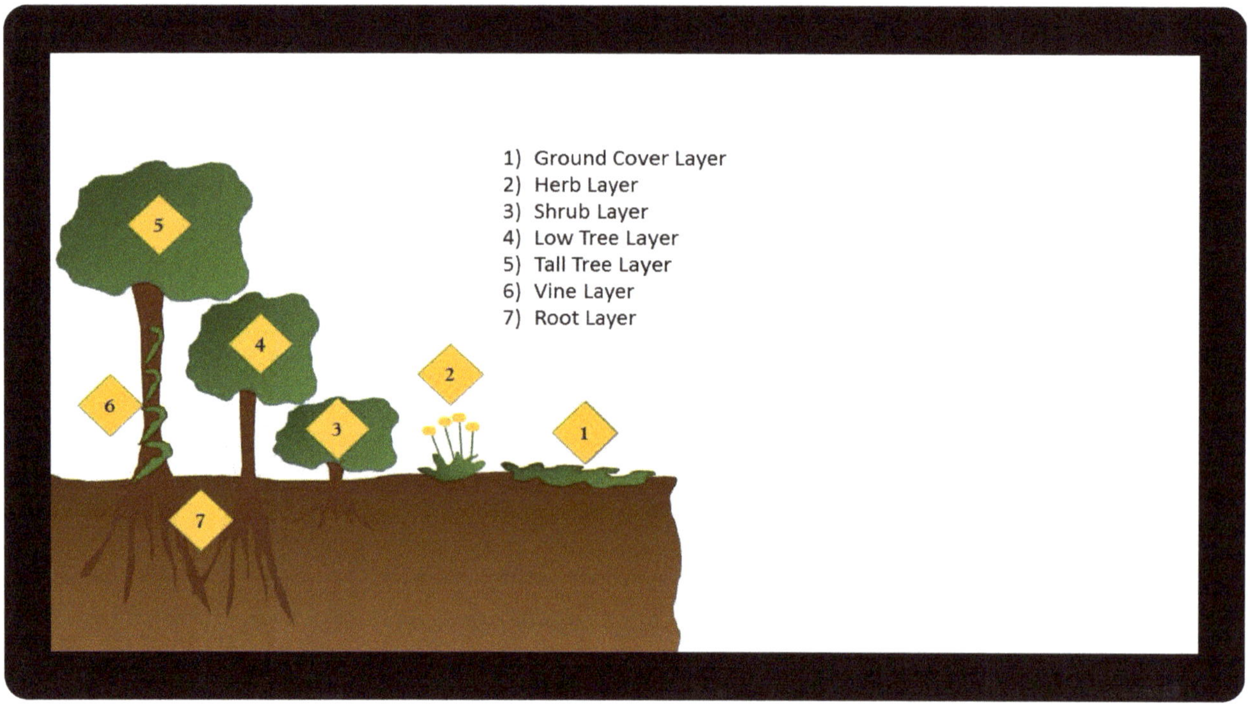

LAYERS OF A FOOD FOREST GARDEN ECOSYSTEM:

- *Ground Cover Layer*: Low, ground–hugging plants that offer food or shelter. These plants do not take up much vertical space and they help immensely with weed control. Examples: strawberries, nasturtium, and creeping thyme.
- *Herb Layer*: In this case, herb is going to stand for herbaceous meaning "nonwoody" plants. This could be vegetables, herbs, edible flowers or even cover crops. Examples: basil, lettuce, squash.
- *Shrub Layer*: This layer contains flowering and fruiting shrubs as well as shrubs that provide habitats for native wildlife. Examples: raspberry, blueberry, and tomato.
- *Low Tree Layer*: This layer has small trees. Fruit trees are a great fit for this layer. Examples: dwarf plum, peach, citrus.
- *Tall Tree Layer*: Make sure that you do not use trees large enough to cast a shadow over the entire garden. Examples: apple, pear and cherry trees.
- *Vine Layer*: This layer is for climbing plants. Examples: grapes, cucumbers, pole beans.
- **Root Layer**: This layer lets us utilize the soil beneath the edible ecosystem. Shallow rooted plants work the best in order to avoid disturbing other plants when you harvest the root vegetables. Examples: potato, onion, garlic.

EDIBLE ECOSYSTEM Name ______________________________

Part I: List three benefits of an edible ecosystem:

- ____
- ____
- ____

Part II: Label the layers of the edible ecosystem below. Write a description for each layer.

1) ____

2) ____

3) ____

4) ____

5) ____

6) ____

7) ____

<u>DESIGN AN EDIBLE ECOSYSTEM</u> Name______________________________

Instructions: Draw and label an edible ecosystem using the plant suggestions from this lesson.

DESIGN AN EDIBLE ECOSYSTEM RUBRIC

Criteria	1	2	3	4
Ecosystem Layers	4 or less layers	5 layers	6 layers	7 layers
Label and Description	Missing most labels and descriptions	Some labels and some descriptions are correct	All labels and most descriptions are correct.	All labels and descriptions are correct.
Creativity	Design lacks color and details	Design is adequate and colorful	Design is nicely drawn but lacks creativity and a unique personal touch	Design is colorful, unique, creative and shows a personal touch

Points scored on rubric	Grade
12	100
11	95
10	90
9	85
8	80
7	75
6	70
5 or below	65

SEVENTH GRADE LESSON PLAN
Garden Genetics

SUBJECT / GRADE LEVEL: 7th Grade - Science.

ACADEMIC STANDARDS AND PERFORMANCE INDICATORS:

Science:

7.L.4 The student will demonstrate an understanding of how genetic information is transferred from parent to offspring and how environmental factors and the use of technologies influence the transfer of genetic information.

7.L.4A.2 Construct explanations for how genetic information is transferred from parent to offspring in organisms that reproduce sexually.

7.L.4A.3 Develop and use models Punnett squares to describe and predict patterns of the inheritance of single genetic traits from parent to offspring including dominant and recessive traits, incomplete dominance, and co-dominance.

LESSON OBJECTIVES:

TSW analyze the transfer of genetic information in plants. TSW develop and use models to predict genetic traits.

MATERIALS:

- *Garden Genetics* student worksheet (attached)
- *Garden Genetics* teacher guide (attached)
- *Garden Genetics Punnett Square Exercise* (attached)

ENGAGEMENT:

- **What is a seed bank?**

EXPLORATION:

- The teacher will begin by explaining the importance of ensuring the genetic diversity of our food supply through seed banks.

EXPLANATION:

- The teacher will follow up the engagement question with the video titled *The Doomsday Vault: Secure your Seed* by Peter Lustig: https://www.youtube.com/watch?v=gFAcyJxIguA.
- Using the *Garden Genetics* Teacher Guide, the teacher will guide students through the completion of the *Garden Genetics* student worksheet.
- The teacher will show the following video to introduce the *Garden Genetics Punnett Square* exercise.
- *How Mendel's pea plants helped us understand genetics* by TedEx: https://www.youtube.com/watch?v=Mehz–tCxjSE.
- The teacher will explain and guide students through the *Garden Genetics Punnett Square* exercise.

ELABORATION:

- If the teacher wishes to extend the lesson, the class can create its own seed bank by starting a seed library. Additionally, students could investigate the lack of genetic diversity that caused the Irish Potato Famine.

EVALUATION:

- Students will turn in the *Garden Genetics* student worksheet and the *Garden Genetics Punnett Square* exercise.

RESOURCES:

1. *The Doomsday Vault: Secure your Seed* by Peter Lustig: https://www.youtube.com/watch?v=gFAcyJxIguA.
2. *How Mendel's pea plants helped us understand genetics* by TedEx https://www.youtube.com/watch?v=Mehz–tCxjSE.
3. General Resource: http://www.fao.org/home/en/.
4. *Garden Genetics* teacher guide (attached).
5. *Garden Genetics* student worksheet (attached).
6. *Garden Genetics Punnett Square* exercise (attached).

TEACHER GUIDE
GARDEN GENETICS

BACKGROUND:

Students should understand that maintaining genetic diversity in our food system guards us against the collapse of our world's food supply due catastrophes such as diseases or natural disasters.

ENGAGEMENT:

- Teacher will pose the questions, "*What is a seed bank?*"

EXPLANATION:

Guide students through the completion of *Garden Genetics* using information below:

- *Seed Banks*: A store seeds in climate controlled facilities often underground in order to protect the seeds from extreme weather or even civil unrest. The seeds are usually dried and kept frozen as well. Some seeds last for many years but others must be replaced to ensure their viability.
- *Agrobiodiversity*: Is extremely important in order to maintain a wide gene pool so that we can continue to breed plants to ensure resistance to diseases as they arise. The diversity of our world's food supply is currently at risk.
- *Svalbard Global Seed Vault*: Is the most famous seed bank located on a frozen island in Norway. Because of the natural permafrost, it has ideal storage conditions for the seeds. There is also no human activity in this region which protects the seeds as well.
- *Community Seed Bank or Library*: Different gardeners in a community saved the seeds from their non–hybrid plants based on their traits for disease resistance or good flavor. Other gardeners can then borrow seeds from the local community seed bank to ensure that we do not let any plant varieties go extinct. At the end of the growing season, the seeds are once again collected from that season's crops and returned to the seed bank.
- *Heirloom Vegetables*: These vegetables come from seed that gets passed down through generations thus ensuring the plant will not go extinct. Heirloom varieties are considered to best "the best of the best" that gardeners have selected to pass down because of a certain trait, often flavor. These vegetables are open-pollinated, which means they pollinate without human help typically by insects or wind. These varieties are typically no less than fifty years old and their characteristics remain stable and get passed down from year to year.

- *Hybrid Vegetables*: Are created when humans intentionally cross–pollinate two different varieties of a plant. The goal is to create a hybrid that contains the best traits of each of the parent plant. Cross-pollination can happen naturally in the wild within members of the same plant species. However, plant breeders carefully control pollination when creating plant hybrids. This process can take several years. Typically the desired characteristics for hybridizing are disease or insect resistance, flavor, size, or post-harvest shelf life. The characteristics of hybrid vegetables do not remain stable from year to year like those of the heirloom varieties. Typically these seeds revert back to one of the parents rather than the combination of both. Therefore, hybrids are not nearly as reliable for seed saving from year to year.

AGROBIODIVERSITY FACTS:

Direct students to the following website to answer the *Garden Genetics* Part II questions: http://www.fao.org/home/en/.

- *How much plant genetic diversity has been lost since the early 1900s?* 75% of genetic plant diversity has been lost worldwide as farmers have left their genetically diverse local plant varieties and have sought out genetically uniform, high yielding hybrid varieties.
- *How many different plant species currently provide 75% of the world's food supply?* Only 12 plant species.
- *Approximately how many edible species of plants exist?* 250,000 to 300,000 known edible plant species.
- *How many of these species to we actually consume?* Only 150 to 200 species are used by humans.
- *Which three species contribute nearly 60% of all calories and proteins consumed by humans on Earth?* Rice, maize and wheat.

PUNNETT SQUARES

- *Genotype*: The set of genes that is responsible for a particular trait. (In the example used for this lesson that trait is the color of peas green or purple).
- *Phenotype*: The physical expression of the genetic trait. (In the example used for this lesson that trait is the color of peas green or purple).

****Teacher will guide students through Punnett Square example then the students will independently complete their own Punnett Squares in the* Garden Genetics Punnett Square *exercise.*

GARDEN GENETICS

Name ______________________________

Part I: *Define the following terms:*

Term	Definition
Seed Banks	
Agrobiodiversity	
Svalbard Global Seed Vault	
Community Seed Bank or Library	
Heirloom Vegetables	
Hybrid Vegetables	

Part II: *Answer the following questions regarding Agrobiodiversity:*

1) How much plant genetic diversity has been lost since the early 1900s?

   ____

2) How many different plant species currently provide 75% of the world's food supply?

   ____

3) Approximately how many edible species of plants exist?

   ____

4) How many of these species to we actually consume?

   ____

5) Which three species contribute nearly sixty percent of all calories and proteins consumed by humans on Earth?

   ____

PUNNETT SQUARES   Name______________________________

Part I: *Define Genotype and Phenotype then see the Punnett Square example below:*

- **Genotype**

  ____

  ____

- **Phenotype**

  ____

  ____

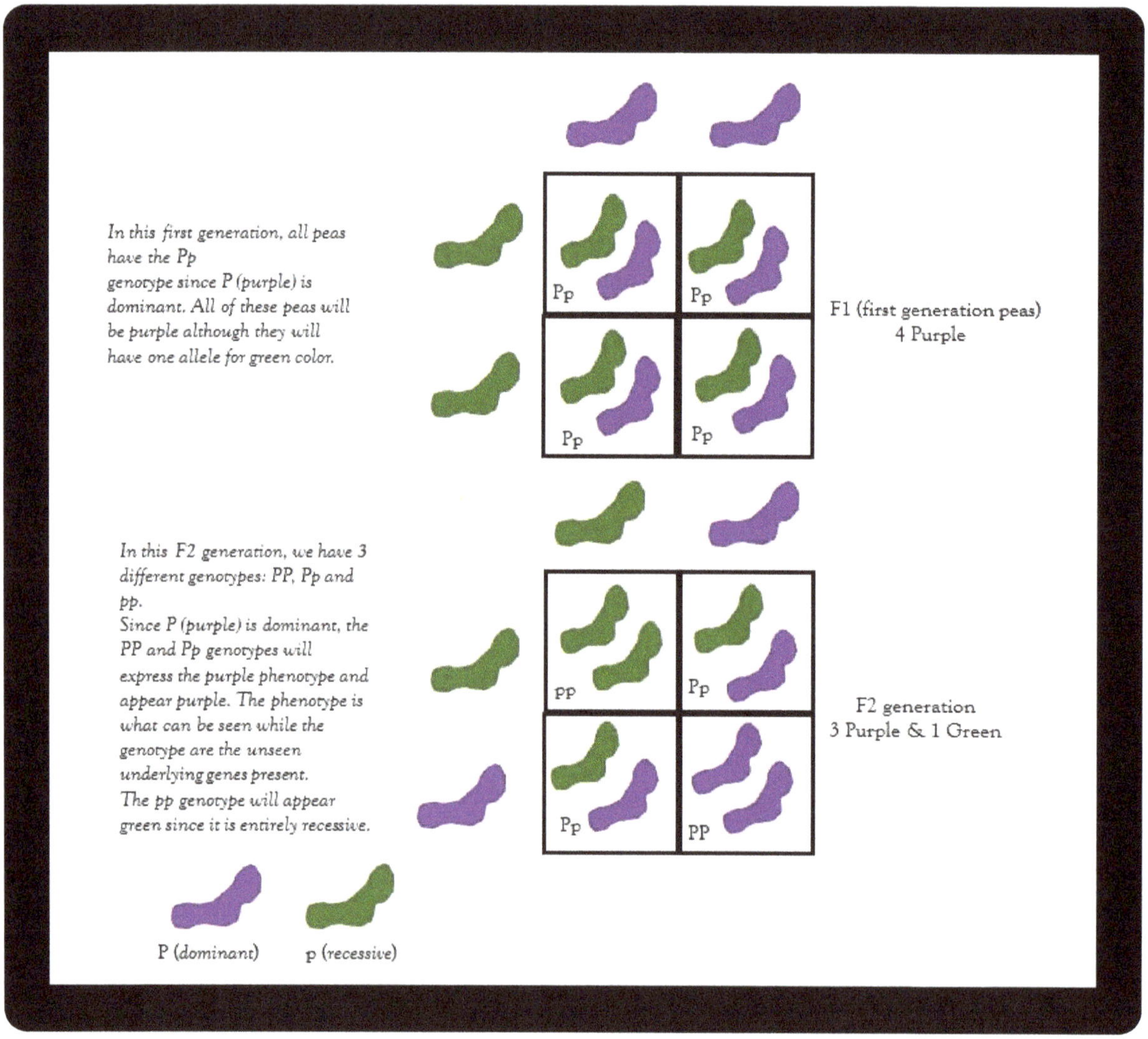

Part II: *Complete the following Punnett Squares following the instructions below:*

	p	P
P		
P		

	P	p
p		
P		

	p	p
P		
P		

	P	p
p		
p		

Based on the information below, complete these Punnett Squares and write in each square the correct results as well as the pea color.

P (*dominant purple*)   p (*recessive green*)

SEVENTH GRADE LESSON PLAN
Grow Up

SUBJECT / GRADE LEVEL: 7th Grade - Science.

ACADEMIC STANDARDS AND PERFORMANCE INDICATORS:

Science:

7.EC.5 The student will demonstrate an understanding of how organisms interact with and respond to the biotic and abiotic components of their environments.

7.EC.5A In all ecosystems, organisms and populations of organisms depend on their environmental interactions with other living things biotic factors and with physical abiotic factors such as light, temperature, water, or soil quality. Disruptions to any component of an ecosystem can lead to shifts in its diversity and abundance of populations.

LESSON OBJECTIVES:

TSW identify the environmental factors that affect plant growth. TSW create a planting chart that will serve as a guide to proper planting times based on optimal temperatures for plant growth.

MATERIALS:

- *Grow Up* student worksheet (attached)
- *Grow Up* teacher guide (attached)

ENGAGEMENT:

- **How do you commit a plant murder?**

EXPLORATION:

- The teacher will begin by exploring the environmental factors that affect plant growth.

EXPLANATION:

- The teacher will follow up the engagement question with the engaging video titled *Light Reactions* by Biology Videos: https://www.youtube.com/watch?v=eDA8rmUP5ZM.

- This time lapse video illustrates the plants ability to recover from lack of water. Using the *Grow Up* teacher guide, the teacher will guide students through the completion of the *Grow Up* student worksheet.

ELABORATION:

- If the teacher wishes to extend the lesson, the class can implement a crop schedule based on the planting chart created on the *Grow Up* student worksheet.

EVALUATION:

- Students will turn in the *Grow Up* student worksheet.

RESOURCES:

1. *Light Reactions* by Biology Videos: https://www.youtube.com/watch?v=eDA8rmUP5ZM.
2. Clemson Extension: https://www.clemson.edu/extension/hgic/plants/vegetables/gardening/hgic1256.html.
3. *Grow Up* teacher guide (attached).
4. *Grow Up* student worksheet (attached).

TEACHER GUIDE
GROW UP

BACKGROUND:

Students should understand that the success or failure of plant growth is based on the interaction between the plant and abiotic factors such as light, temperature, water and nutrient availability.

ENGAGEMENT:

- Teacher will pose the question, "**How do you commit a plant murder?**" In other words, how do you kill a plant—intentionally or not?

EXPLANATION:

ENVIRONMENTAL FACTORS THAT AFFECT PLANT GROWTH:

- WATER: Water amount and quality are both important to ensure proper plant growth. When plants are suffering from lack of water their leaves will exhibit wilting. As long as you provide them with water at that point, they will recover. If you do not then they will reach their permanent wilting point in which they will die. It is important to know that too much water can also kill plants. Wilting can also be a sign of too much water so make sure you check the soil for moisture before adding water to an already wilted plant.
- LIGHT: Plants need both the proper amount of light and the correct intensity. Most plants like full sun but others will scorch without the protection of partial shade. Light intensity changes with the seasons. Winter days are shorter and the light is less intense. The intense sunlight of the summer promotes flowering and fruiting of most plants.
- TEMPERATURE: Plants react to temperature by speeding up or slowing down processes such as respiration, transpiration, and photosynthesis. Plants grow more quickly during the warm season and slow down sometimes becoming dormant during the cool season. Warm–season vegetables and most flowers grow best between 60° and 80° F. Cool–season vegetables grow best between 50° and 70° F. Temperatures out of these optimum ranges can result in abnormal development and reduced fruit and vegetable production.
- *Where does soil fit into this equation?* Plants do not actually need soil to grow as evidenced with hydroponics. However, even with hydroponics, plants must have the proper nutrients for plant growth. We can add nutrients to the growing medium in order to maximize plant growth and health for each particular food crop.

****Guide students through the completion of Part II of the* Grow Up *student worksheet using the following website for cool and warm season crop suggestions:*
https://www.clemson.edu/extension/hgic/plants/vegetables/gardening/hgic1256.html.

Grow Up!

Name ______________________________

Part I: Complete the chart below describing some of the most important environmental factors that affect plant growth.

Water	
Light	
Temperature	

Part II: Create a planting chart based on the Spring and Fall planting dates as well as your location.

Crop	Days to Harvest	Planting Depth (*inches*)	Spring Planting Dates	Fall Planting Dates

SEVENTH GRADE LESSON PLAN

Invasive Species

SUBJECT / GRADE LEVEL: 7th Grade - Science.

ACADEMIC STANDARDS AND PERFORMANCE INDICATORS:

Science:

7.EC.5 The student will demonstrate an understanding of how organisms interact with and respond to the biotic and abiotic components of their environments.

7.EC.5B.4 Define problems caused by the introduction of a new species in an environment and design devices or solutions to minimize the impact s to the balance of an ecosystem.

LESSON OBJECTIVES:

TSW identify and define the problems created when an invasive plant species is introduced into an ecosystem. TSW investigate measures to keep invasive species under control.

MATERIALS:

- *Invasive Species* student worksheet (attached)
- *Invasive Species* teacher guide (attached)

ENGAGEMENT:

- **What does the word *invasive* mean?**

EXPLORATION:

- The teacher will ask students if they know of an invasive plant or animal species. This will then lead into the discussion of invasive or exotic species taking over our native plant and animal species.

EXPLANATION:

- The teacher will follow up the exploration discussion with the engaging video titled *Invasive Species: The Story of Bunny* by SciShow: https://www.youtube.com/watch?v=eDOwTXobJ3k.

- Using the *Invasive Species* teacher guide, the teacher will guide students through the completion of *Invasive Species* student worksheet.

ELABORATION:

- If the teacher wants to extend the lesson, the class can implement a campus-wide project to identify and eradicate invasive plant species.

EVALUATION:

- Students will turn in the *Invasive Species* student worksheet.

RESOURCES:

1. *Invasive Species: The Story of Bunny* by SciShow: https://www.youtube.com/watch?v=eDOwTXobJ3k.
2. General Resource: http://www.usna.usda.gov/Gardens/invasives.html.
3. *Invasive Species* teacher guide (attached).
4. *Invasive Species* student worksheet (attached).

TEACHER GUIDE
INVASIVE SPECIES

BACKGROUND:

Students should understand that invasive species can adapt quickly to a new environment and reproduce rapidly, often causing a decline in native species.

ENGAGEMENT:

- Teacher will ask students, "**What does the word invasive mean?**"

EXPLANATION:

Guide students through the completion of the *Invasive Species* student worksheet using the *Invasive Species: The Story of Bunny* by SciShow video as well as the descriptions below:

- Invasive species: Are organisms that adapt quickly to new environments by reproducing and spreading rapidly, often displacing the organisms that were originally there.
- Exotic species: Are organisms that are not native to an area.
  –Not all exotic species are invasive.
  –Many crops that we grow in the United States are exotic. These plants are only an issue if they escape human cultivation and begin to spread rapidly in the wild competing with native species.
- The cost of invasive species: In North America, it is estimated that there are about 50,000 species of exotic plants, animals, and pathogens currently established. These invasive species cost our economy an close to $138 billion per year.
  –Don't bring back plants or seeds from trips to other countries.
  –Learn to identify species invasive to your area. Don't plant them and remove them when possible.

INVASIVE SPECIES EXAMPLES

Common Name	Botanical Name	Introduced	Description	Spreads
Chinese Tallow Tree	Triadica sebifera	From China in the late 1700s.	Deciduous tree that grows to 60 feet in height.	Seeds dispersed by birds and water.
Chinese Privet	Ligustrum sinense	Introduced from China in the early to mid 1800s.	Semi–evergreen, thicket forming shrub growing to 30 feet in height.	Spreads by abundant bird and animal dispersal and root sprouts.
Beach Vitex	Vitex rotundifolia	Introduced from Korea in the mid 1980s as an ornamental and for dune stabilization.	Groundcover vine with round, silvery, gray-green leaves.	Seeds are spread by animals, wind and water.
Japanese Honeysuckle	Lonicera japonica	Introduced from Japan in the 1800s.	Semi-evergreen to evergreen woody vine, is high climbing and trailing to 80 feet.	Spreads by rooting at vine nodes and animal dispersal of seed.

INVASIVE SPECIES

Name ______________________________

Instructions: Answer the following questions and then complete the invasive species example chart below.

1) What is an invasive species?

____

____

2) What is an exotic species?

____

____

3) Are all exotic species native?

____

4) How do we keep exotic species from becoming invasive?

____

____

5) How many exotic species are estimated to be in North America?

____

6) How much do these invasive species cost us each year?

____

7) How can you prevent the spread of invasive species?

____

____

____

Invasive Species Examples

Common Name	**Botanical Name**	**Introduced**	**Description**	**Spreads**

SEVENTH GRADE LESSON PLAN
Lost Without Leaves

SUBJECT / GRADE LEVEL: 7th Grade - Science - Duration: Two sixty minute lessons.

ACADEMIC STANDARDS AND PERFORMANCE INDICATORS:

Science:

7.L.3 The student will demonstrate an understanding of how the levels of organization within organisms support the essential functions of life.

7.L.3A.1 Obtain and communicate information to support claims that 1.) organisms are made of one or more cells, 2.) cells are the basic unit of structure and function of organisms, and 3.) cells come only from existing cells.

7.L.3A.2 Analyze and interpret data from observations to describe different types of cells and classify cells as plant, animal, protist, or bacteria.

LESSON OBJECTIVES:

TSW identify the structures within plant cells that enable them to conduct photosynthesis. TSW compare the modified leaves that allow plants to survive and thrive.

MATERIALS:

- *Lost Without Leaves* student worksheet (attached)
- *Lost Without Leaves* teacher guide (attached)
- *Leaf Collection* worksheet (attached)

ENGAGEMENT:

- Compare and contrast a leaf and a solar panel.

EXPLORATION:

- DAY ONE: The teacher will begin by creating a Venn Diagram that compares a leaf and a solar panel.
- DAY TWO: Students will complete *Leaf Collection* worksheet.

EXPLANATION:

- DAY ONE: Using the *Lost Without Leaves* teacher guide, the teacher will guide students through the completion of the *Lost Without Leaves* student worksheet.
- The teacher will encourage students to further explore the different types of modified leaves for homework to be shared with the class on the following day.
- DAY TWO: Students will complete *Leaf Collection* worksheet.

ELABORATION:

- If the teacher wishes to extend the lesson, he or she can collect plants with modified leaves, such as insectivorous or reproductive, to keep in the classroom.

EVALUATION:

- DAY ONE: Students will turn in *Lost Without Leaves* student worksheet.
- DAY TWO: Students will turn in *Leaf Collection* worksheet.

RESOURCES:

1. Garden Notes: http://www.ext.colostate.edu/mg/gardennotes/134.html.
2. Leaf Resource: http://www.uky.edu/Ag/Horticulture/Geneve/teaching/PLS%20220/Leafarrangement01.pdf.
3. *Lost Without Leaves* teacher guide (attached).
4. *Lost Without Leaves* student worksheet (attached).
5. *Leaf Collection* worksheet (attached).

TEACHER GUIDE
LOST WITHOUT LEAVES

BACKGROUND:

Students should understand that photosynthesis is essential to life on earth. Leaves has adapted to modify their structure in order to survive and thrive.

ENGAGEMENT:

- Ask students to complete the Venn Diagram on their comparing and contrasting a leaf and a solar panel.

EXPLANATION:

Guide students through the completion of the *Lost Without Leaves* student worksheet using the information below:

LEAF STRUCTURES:

- Guide students through the completion of the *Lost Without Leaves* student worksheet using the following website: http://www.ext.colostate.edu/mg/gardennotes/134.html.

MODIFIED LEAVES:

- *Insectivorous leaves*: Insectivorous leaves often grow on plants in acidic swamps which are deficient in nutrients needed by the plants. Their nutritional needs are met by the absorption of nutrients from insects.
- *Reproductive leaves*: Tiny but complete plantlets along the edge of the leaves. Each plantlet, when separated from the leaf, is capable of growing independently into a full-sized plant.
- *Bracts*: Showy, often colorful, modified leaves that surround the true flowers to attract pollinators. Poinsettias and dogwoods have bracts. The actual flowers are small, greenish-yellow, and located in the center of the bracts.
- *Spines*: These modified leaves are often found on desert dwellers such as cactus. Spines reduce the surface area of the leaf which reduces water loss and can offer defense against predators. Spines should not be confused with thorns which are actually a modified type of stem.
- *Shade leaves*: Leaves can actually increase their surface area and grow larger in order to collect more sunlight when they are growing in low light situations. This modification is remarkable, as both types of leaves on the plant have exactly the same genes. The shade leaves are just larger in size in order to collect more sunlight.

LEAF COLLECTION:

- See charts on the following website for the Leaf Collection exercise: http://www.uky.edu/Ag/Horticulture/Geneve/teaching/PLS%20220/Leafarrangement01.pdf.

LOST WITHOUT LEAVES Name________________________________

Instructions: Complete the Venn Diagram below comparing and contrasting a leaf and a solar panel.

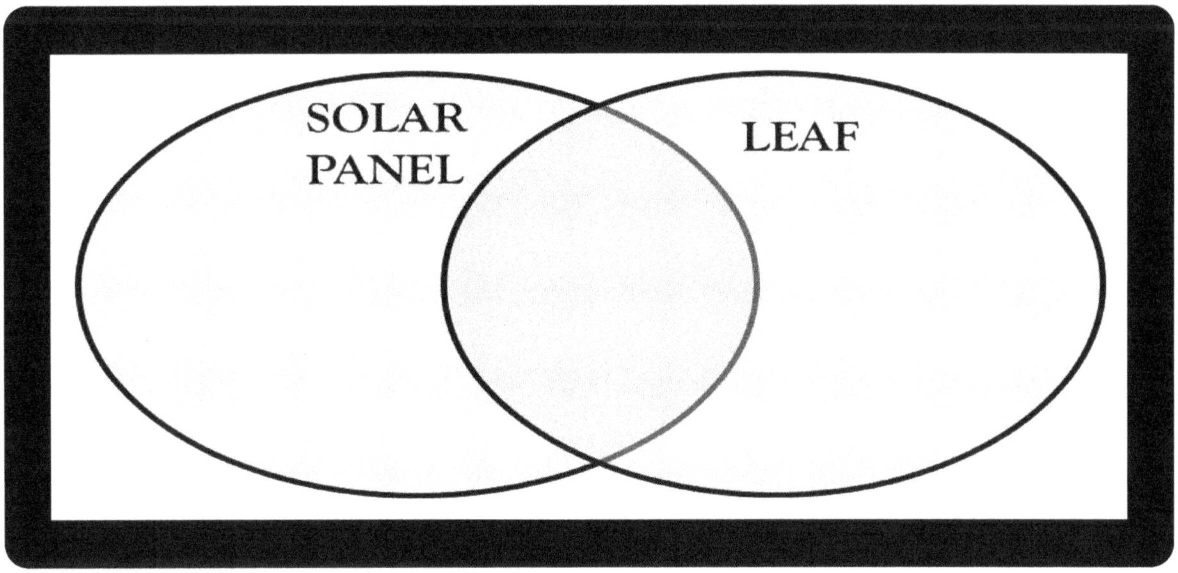

Define the following internal structures of leaves:

Epidermis-

Palisade layer-

Chloroplasts-

Spongy mesophyll-

Stomata-

Guard cells-

Modified Leaf	Description

Leaf Collection

Name________________________________

Instructions: Sketch a leaf arrangement in each box below. In the adjoining box, tape or glue an actual leaf arrangement that corresponds with your drawing.

Leaf Arrangement Sketch	Actual Leaf Arrangement Sample

SEVENTH GRADE LESSON PLAN
Mighty Microbes

SUBJECT / GRADE LEVEL: 7th Grade - Science - Duration: Two sixty minute lessons.

ACADEMIC STANDARDS AND PERFORMANCE INDICATORS:

Science:

7.L.3A.4 Construct scientific arguments to support claims that bacteria are both helpful and harmful to other organisms and their environment.

LESSON OBJECTIVES:

TSW compare harmful and beneficial bacteria in the garden. TSW create a garden pest management plan utilizing commercially available beneficial bacteria.

MATERIALS:

- *Mighty Microbes* teacher guide (attached)
- *Mighty Microbes* student worksheet (attached)
- *Garden Pest Management Plan* (attached)

ENGAGEMENT:

- **What is the five second rule?**

EXPLORATION:

- The teacher will begin by explaining that bacteria can be harmful or helpful.

EXPLANATION:

- The teacher will follow up the engagement question with the video titled *Testing the 5 Second Rule* by RMIT University: https://www.youtube.com/watch?v=dl5k3RxhQG8.
- Using the *Mighty Microbes* teacher guide, the teacher will guide students through the completion the *Mighty Microbes* student worksheet.
- Students will complete the *Garden Pest Management Plan.*

EVALUATION:

- Students will turn in the *Mighty Microbes* student worksheet as well as the *Garden Pest Management Plan.*

ELABORATION:

- If the teacher wishes to extend the lesson, the class can implement the *Garden Pest Management Plan* utilizing commercially available beneficial bacteria.

RESOURCES:

1. *Testing the 5 Second Rule* by RMIT University: https://www.youtube.com/watch?v=dl5k3RxhQG8.
2. Beneficial Organisms: http://www.arbico-organics.com/category/Beneficial-Organisms.
3. General Resource: http://www.microbeworld.org/.
4. *Mighty Microbes* teacher guide (attached).
5. *Mighty Microbes* student worksheet (attached).
6. *Garden Pest Management Plan (*attached).

TEACHER GUIDE
MIGHTY MICROBES

BACKGROUND:

Students should understand that bacteria can be harmful in the case of bacterial diseases but useful as well in the case of probiotics and other beneficial bacteria.

ENGAGEMENT:

- Teacher will pose the question, "**What is the five second rule?**"

EXPLANATION:

- Teacher will explain the risk of harmful bacteria exposure through the following video: *Testing the 5 Second Rule* by RMIT University: https://www.youtube.com/watch?v=dl5k3RxhQG8.
- Teacher will guide students through the completion of the *Mighty Microbes* student worksheet using the information below:

MICROBES FUN FACTS:

- They are the oldest life form on our planet. They are single celled and so small that millions of them could fit into the end of a pencil.
- There are more microbes on your hand right now than there are people on the planet.
- They could survive without us but without them we could not eat or even breathe.
- Plants couldn't grow without them and garbage would never decay!
- Without microbes, we couldn't digest our food and there would be a lot less oxygen to breathe.

TYPES OF MICROBES:

- Use the following website to identify different types of microbes as well as their important jobs: http://www.microbeworld.org/.

USING MICROBES IN YOUR GARDEN:

- Use the following website to create a garden pest management plan utilizing commercially available microorganisms. http://www.arbico-organics.com/category/beneficial-insects-organisms.

MIGHTY MICROBES Name______________________________

Part I: List five "fun facts" about microbes:

1) ____
2) ____
3) ____
4) ____
5) ____

Part II: Complete the following chart identifying different types of microbes as well as their important jobs.

Microbe	What It Does

GARDEN PEST MANAGEMENT PLAN Name________________________

Instructions: Create a garden pest management plan utilizing commercially available microorganisms.

Beneficial Microrganism/Product	Pest(s) Controlled

SEVENTH GRADE LESSON PLAN

Soggy Soils

SUBJECT / GRADE LEVEL: 7th Grade - Science.

ACADEMIC STANDARDS AND PERFORMANCE INDICATORS:

Science:

7.EC.5 The student will demonstrate an understanding of how organisms interact with and respond to the biotic and abiotic components of their environments.

7.EC.5A.2 Construct explanations of how soil quality including composition, texture, particle size, permeability, and pH affects the characteristics of an ecosystem using evidence from soil profiles.

LESSON OBJECTIVES:

TSW investigate the water holding capacity of soil. TSW understand that small soil particle size enables soil to hold water whereas large particles drain water very quickly.

MATERIALS:

- *Soggy Soils* student worksheet (attached)
- *Soggy Soils* teacher guide (attached)

ENGAGEMENT:

- **How can you increase the water holding capacity of the soil in your garden?**

EXPLORATION:

- The teacher will begin lesson by explaining that increasing the water holding capacity of your garden can result in healthier and happier plants during the course of the growing season.

EXPLANATION:

- The teacher will follow up the engagement question with the video titled *Soil Composition and Water Holding Capacity* by Oklahoma Gardening: https://www.youtube.com/watch?v=xO48IbeuiPM.

- Using the *Soggy Soils* teacher guide, the teacher will guide students through the completion of *Soggy Soils* student worksheet.

EVALUATION:

- Students will turn in the *Soggy Soils* student worksheet.

ELABORATION:

- If the teacher wishes to extend the lesson, the class can set up a compost bin. By adding more organic matter to the garden, the students will increase the water holding capacity of the soil.

RESOURCES:

1. General Resource: http://msue.anr.msu.edu/news/why_is_soil_water_holding_capacity_important.
2. *Soil Composition and Water Holding Capacity* by Oklahoma Gardening: https://www.youtube.com/watch?v=xO48IbeuiPM.
3. *Soggy Soils* teacher guide (attached).
4. *Soggy Soils* student worksheet (attached).

TEACHER GUIDE
SOGGY SOILS

BACKGROUND:

Students should understand that the water holding capacity of soils is affected by the size of soil particles.

ENGAGEMENT:

- Teacher will pose the question, "**How can you increase the water holding capacity of the soil in your garden?**"

EXPLANATION:

- Teacher will follow up the engagement question with the video titled: *Soil Composition and Water Holding Capacity* by Oklahoma Gardening: https://www.youtube.com/watch?v=xO48IbeuiPM.
- Teacher will guide students through the completion of the *Soggy Soils* student worksheet using the information found on the following website: http://msue.anr.msu.edu/news/why_is_soil_water_holding_capacity_important.

***SOGGY SOILS* KEY:**

- Soil Water Holding Capacity: Is the amount of water that a given soil can hold for crop use.
- Field Capacity: Is the point where the soil water holding capacity has reached its maximum for the entire field.
- Soil Texture: Refers to the weight proportion (relative proportion by weight percentage of sand, silt, and clay) of the mineral soil separates for particles less than two millimeters (mm) as determined from a laboratory particle-size distribution.
- Soil Organic Matter: (SOM) is the organic matter component of soil, consisting of plant and animal residues at various stages of decomposition, cells and tissues of soil organisms, and substances synthesized by soil organisms.
- *What can you do to increase the soil watering holding capacity in your garden?* Add organic matter.
- *List four ways to increase organic matter in the soil:*
  1. Leaves from deciduous trees can be left in place to decompose.
  2. Plant debris can be composted and incorporated back into gardens.
  3. Plant residue and animal manures can be incorporated directly into the soil.
  4. Plant cover crops.

SOGGY SOILS Name ______________________________

Instructions: Define the following terms as well as answer the last two questions.

Term	Definition
Soil Water Holding Capacity	
Field Capacity	
Soil Texture	
Soil Organic Matter	

1) What can you do to increase the soil watering holding capacity in your garden?

2) List four ways to increase organic matter in the soil:

    a)

    b)

    c)

    d)

SEVENTH GRADE LESSON PLAN
Waiting On The Water Cycle

SUBJECT / GRADE LEVEL: 7th Grade - Science - Duration: Two sixty minute lessons.

ACADEMIC STANDARDS AND PERFORMANCE INDICATORS:

Science:

7.EC.5 The student will demonstrate an understanding of how organisms interact with and respond to the biotic and abiotic components of their environments.

7.EC.5A In all ecosystems, organisms and populations of organisms depend on their environmental interactions with other living things biotic factors and with physical abiotic factors such as light, temperature, water, or soil quality. Disruptions to any component of an ecosystem can lead to shifts in its diversity and abundance of populations.

LESSON OBJECTIVES:

TSW differentiate between the components of the water cycle which provides precipitation for plant use. TSW recognize that rain can be collected and stored for future use.

MATERIALS:

- *Waiting On The Water Cycle* student worksheet (attached)
- *Waiting On The Water Cycle* teacher guide (attached)
- *Design A Rain Barrel* student worksheet (attached)

ENGAGEMENT:

- DAY ONE: **When garden needs to be watered, how can you make it rain?**
- DAY TWO: **Since we can t force it to rain how can we capture rainwater in order collect and save it?**

EXPLORATION:

- DAY ONE: The teacher will begin by discussing the water cycle and the origin of rain.
- DAY TWO: The teacher will ask students to list ways in which rainwater can be captured for reuse.

EXPLANATION:

- DAY ONE: The teacher will follow up the rain question with the video titled *The Water Cycle* by National Science Foundation: https://www.youtube.com/watch?v=al-do-HGuIk.
- Using the *Waiting On The Water Cycle* teacher guide, the teacher will guide students through the completion of the *Waiting On The Water Cycle* student worksheet.
- DAY TWO: Teacher will show students the following video describing the process of building a rain barrel:
- *Building a Rain Barrel - Easy as 1, 2, 3* by Okanagan WaterWise: https://www.youtube.com/watch?v=K2qoLtvTYN0.
- The teacher will then divide students into pairs. Each pair will conduct a site assessment to determine the best location for a school garden rain barrel at their school. Students will complete the *Design a Rain Barrel* student worksheet.

EVALUATION:

- DAY ONE: Students will turn in the *Waiting On The Water Cycle* student worksheet.
- DAY TWO: Students will turn in the *Design A Rain Barrel* student worksheet.

ELABORATION:

- If the teacher wishes to extend the lesson, the class can install a rain barrel to collect water for the school garden.

RESOURCES:

1. *The Water Cycle* by National Science Foundation: https://www.youtube.com/watch?v=al-do-HGuIk.
2. General Resource: http://www.cbf.org/Document.Doc?id=30.
3. *Building a Rain Barrel - Easy as 1, 2, 3* by Okanagan WaterWise: https://www.youtube.com/watch?v=K2qoLtvTYN0.
4. *Waiting on the Water Cycle* teacher guide (attached).
5. *Waiting on the Water Cycle* student worksheet (attached).
6. *Design a Rain Barrel* student worksheet (attached).

TEACHER GUIDE
WAITING ON THE WATER CYCLE

BACKGROUND:

Students should understand that all of the water in earth is all that there will ever be. It is constantly changing from forms from liquid to vapor to solid. This is called the water cycle and it is powered by the sun.

ENGAGEMENT:

- Teacher will ask students. "**When garden needs to be watered, how can you make it rain?**"
- Teacher will explain that we cannot make it rain. The water cycle determines when it rains. However, on day two of this lesson we will explore ways to capture rain and save it for future use.

EXPLANATION:

- Teacher will guide students through the completion of the Part I of the *Waiting On The Water Cycle* student worksheet using the drawings and descriptions below:

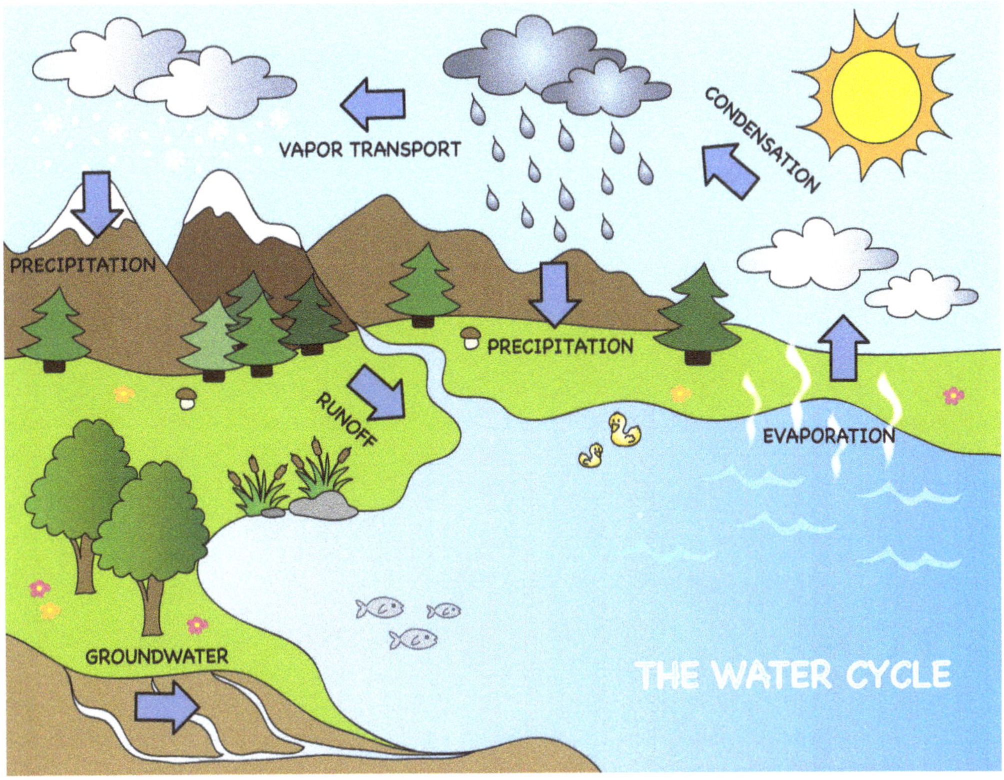

THE WATER CYCLE:

- *Groundwater*: The water that exists below the Earth's surface.
- *Evaporation*: Occurs when the sun heats water up and it becomes water vapor in the air.
- *Condensation*: Occurs when air cools and the water vapor collects in the form of liquid. Condensation is the opposite of evaporation.
- *Transpiration*: The process that carries water through plants from the roots to small pores on the underside of leaves. These pores, called *stomata*, allow water to be released into the air in the form of water vapor.
- *Runoff*: Occurs when precipitation falls faster than it can be absorbed into the ground.
- *Precipitation*: Condensation falling onto land. It can be in the form of mist, rain, sleet, hail, or snow.
- *Solar Radiation*: Radiant energy released by the sun.
- *Infiltration*: The downward movement of water from the surface of the land into the soil. Once precipitation infiltrates the soil, it becomes groundwater.

COLLECTING RAIN WATER:

Teacher will show the Building a *Rain Barrel—Easy as 1, 2, 3* by Okanagan WaterWise video then guide students through the completion of the Part II of the *Waiting On The Water Cycle* student worksheet using the guidelines below for choosing and constructing a rain barrel.

- If you are collecting rain water off of a roof, contaminates can leach into the water from shingled roofing. For your edible garden, consider collecting rainwater from metal roofs or other locations.
- Place your rain barrel on a level surface and preferably a raised platform to increase the water pressure. Four feet is a good height if possible.
- Make sure you include an overflow hose for times of heavy rainfall.
- Include screening under the lid to keep critters out.
- The following website will also offer additional guidelines, supplies, and a materials budget list: http://www.cbf.org/Document.Doc?id=30.

Waiting On The Water Cycle Name ______________________________

Part I: Define the following components of the water cycle:

Runoff	
Groundwater	
Infiltration	
Evaporation	
Solar Radiation	
Condensation	
Precipitation	
Transpiration	

Based on the descriptions above, identify the components of the water cycle below:

Design A Rain Barrel Name(s) ____________________________

Part II: In this exercise, you are going to conduct a site analysis to determine the best location for a rain barrel at your school garden. Using the guidelines outlined in this lesson you will explain why you have chosen this site. In the area below, you will sketch the design and location of the rain barrel.

Describe three potential locations for your rain barrel:

1) ____

2) ____

3) ____

In the chart below, list the potential positive and negative benefits of each potential location:

	Positive	Negative
Location #1		
Location #2		
Location #3		

After analyzing each location, choose one to use for your location design. In the area below, sketch the design and location of your rain barrel.

SEVENTH GRADE LESSON PLAN
Scavenger Hunt

SUBJECT / GRADE LEVEL: 7th Grade - Science.

ACADEMIC STANDARDS AND PERFORMANCE INDICATORS:

Science:

7.EC.5 The student will demonstrate an understanding of how organisms interact with and respond to the biotic and abiotic components of their environments.

7.EC.5B.1 Develop and use models to explain how organisms interact in a competitive or mutually beneficial relationship for food, shelter, or space including competition, mutualism, commensalism, parasitism, and predator-prey relationships.

7.EC.5A.1 Develop and use models to describe the characteristics of the levels of organization within ecosystems including species, populations, communities, ecosystems, and biomes.

7.L.4 The student will demonstrate an understanding of how genetic information is transferred from parent to offspring and how environmental factors and the use of technologies influence the transfer of genetic information.

7.L.4A.2 Construct explanations for how genetic information is transferred from parent to offspring in organisms that reproduce sexually.

7.L.4A.3 Develop and use models Punnett squares to describe and predict patterns of the inheritance of single genetic traits from parent to offspring including dominant and recessive traits, incomplete dominance, and co-dominance.

7.EC.5B.4 Define problems caused by the introduction of a new species in an environment and design devices or solutions to minimize the impact's to the balance of an ecosystem.

7.L.3 The student will demonstrate an understanding of how the levels of organization within organisms support the essential functions of life.

7.L.3A.1 Obtain and communicate information to support claims that 1.) organisms are made of one or more cells, 2.) cells are the basic unit of structure and function of organisms, and 3.) cells come only from existing cells.

7.L.3A.2 Analyze and interpret data from observations to describe different types of cells and classify cells as plant, animal, protist, or bacteria.

7.EC.5A.2 Construct explanations of how soil quality including composition, texture, particle size, permeability, and pH affects the characteristics of an ecosystem using evidence from soil profiles.

LESSON OBJECTIVES:

TSW examine the benefits of mutualism in the garden provided by companion plantings. TSW discover the array of combinations of plants that can be put together in naturalistic patterns that create mutually beneficial relationships creating a garden ecosystem. TSW analyze the transfer of genetic information in plants. TSW identify the environmental factors that affect plant growth. TSW identify and define the problems created when an invasive plant species is introduced into an ecosystem. TSW identify the structures within plant cells that enable them to conduct photosynthesis. TSW compare the modified leaves that allow plants to survive and thrive. TSW understand that small soil particle size enables soil to hold water whereas large particles drain water very quickly.

MATERIALS:

- *Companion Plantings* student worksheet and *Planting Chart*
- *Edible Ecosystem* student worksheet
- *Garden Genetics* student worksheet
- *Grow Up* student worksheet
- *Invasive Species* student worksheet
- *Lost Without Leaves* student worksheet
- *Leaf Collection* worksheet
- *Soggy Soils* student worksheet
- Paper lunch sacks to collect items
- *Scavenger Hunt* student worksheet (attached)

******Students should complete the lessons listed above to gain prior knowledge in order to complete the scavenger hunt.*

ENGAGEMENT:

- **How many plant terms can you name?**

EXPLORATION:

- The teacher will poll students for plant terms until all of the terms for the scavenger hunt have been stated.

EXPLANATION:

- Using the *Scavenger Hunt* chart, the teacher will guide students through the explanation of all terms in the chart.
- After reviewing the *Scavenger Hunt* chart, students will go to the garden in order to find the items.

EVALUATION:

- Students will show teacher items in the bag as the teacher checks them off of the *Scavenger Hunt* chart.

ELABORATION:

- Students can compost all of the items from the scavenger hunt including the chart and the paper bag. If there is nowhere to compost the items, this could also be turned into a digital scavenger hunt where students simply take pictures of the items in the chart.

RESOURCES:

1. *Scavenger Hunt* student worksheet (attached).

SCAVENGER HUNT

Name ______________________________

Instructions: Find all of the items listed below:

Item	Where did you find it?
1. A root vegetable	
2. A legume	
3. A leaf vegetable	
4. One prehistoric plant	
5. A seed pod that exhibits a Fibonacci sequence of numbers	
6. Leaves that exhibit a Fibonacci sequence of numbers	
7. A flower that exhibits a Fibonacci sequence of numbers	
8. A plant that absorbs blue and red wavelengths of light	
9. A sample from the A horizon of the soil profile	
10. Soil Sample with the texture of Sandy Loam or Sandy Clay	

LESSONS FOR EIGTH GRADE

EIGHTH GRADE LESSON PLAN
Carnivorous Plants

SUBJECT / GRADE LEVEL: 8th Grade - Science.

ACADEMIC STANDARDS AND PERFORMANCE INDICATORS:

Science:

8.E.6 The student will demonstrate an understanding of Earth's geologic history and its diversity of life over time.

8.E.6B.1 Construct explanations for how biological adaptations and genetic variations of traits in a population enhance the probability of survival in a particular environment.

LESSON OBJECTIVES:

TSW analyze carnivorous plants to see how their adaptations have allowed them to survive and thrive. TSW investigate the environments in which carnivorous plants have lived and adapted to grow throughout time.

MATERIALS:

- *Carnivorous Plants* student worksheet (attached)
- *Carnivorous Plants* teacher guide (attached)
- *Design A Carnivorous Plant Project and Rubric* (attached)

ENGAGEMENT:

- **Why do carnivorous plants eat bugs?**

EXPLORATION:

- The teacher will begin by explaining to students that pitcher plants and Venus flytraps are two of our native carnivorous plants.

EXPLANATION:

- The teacher will show the three following videos to introduce carnivorous plants:
- *Peter's Savage Garden* by Exploratorium: http://www.exploratorium.edu/tv/index.php?project=18%20program=615.
- *Inside A Pitcher Plant* by Exploratorium: http://www.exploratorium.edu/tv/index.php?project=18%20program=616.
- *Venus Flytrap Feast* by Exploratorium: http://www.exploratorium.edu/tv/index.php?project=18&program=617.

EVALUATION:

- Student will turn in *Carnivorous Plants* student worksheet and *Design A Carnivorous Plant Project.*

ELABORATION:

- An extension to this lesson would be the creation of a rain garden that contained native carnivorous plants.

RESOURCES:

1. *Peter's Savage Garden* by Exploratorium: http://www.exploratorium.edu/tv/index.php?project=18%20program=615.
2. *Inside A Pitcher Plant* by Exploratorium http://www.exploratorium.edu/tv/index.php?project=18%20program=616.
3. *Venus Flytrap Feast* by Exploratorium http://www.exploratorium.edu/tv/index.php?project=18&program=617.
4. Botanical Society of America: https://sites.google.com/site/cypressgardensplantopedia/home/carnivorous-plants.
5. *Cypress Gardens Plantopedia.*
6. *Carnivorous Plant* student worksheet (attached).
7. *Carnivorous Plants* teacher guide (attached).
8. *Design a Carnivorous Plant Project and Rubric* (attached).

TEACHER GUIDE
CARNIVOROUS PLANTS

BACKGROUND:

Students should understand that plants growing in nutrient deficient environments have developed the adaptation to ingest insects in order to obtain the nutrients that they need to grow.

ENGAGEMENT:

- Teacher will pose the question, "**Why do carnivorous plants eat bugs?**"

EXPLANATION:

- Guide students through the completion of the *Carnivorous Plants* student worksheet by directing students to the Botanical Society of America website: http://botany.org/Carnivorous_Plants/.

CARNIVOROUS PLANTS STUDENT WORKSHEET KEY:

1. *Do carnivorous plants eat anything other than insects?* Yes, small frogs and mammals.
2. *Most plants absorb nutrients through their roots, where do carnivorous plants absorb their nutrients?* Through their leaves.
3. *What are the two biggest threats to carnivorous plants in the wild?* Habitat destruction and over collection.
4. *How many carnivorous plant species have been documented?* 600.
5. *List five types of carnivorous plant traps:* Pitfall, Flypaper, Snap, Suction, Lobster pot.

EVALUATION:

- Display the *Cypress Gardens Plantopedia* website for students to complete the carnivorous plant chart: https://sites.google.com/site/cypressgardensplantopedia/home/carnivorous-plants.

CARNIVOROUS PLANTS

Name________________________________

Instructions: Answer the following questions and complete the chart below.

1. Do carnivorous plants eat anything other than insects?
____
____
____

2. Most plants absorb nutrients through their roots, where do carnivorous plants absorb their nutrients? ______________________________

3. What are the two biggest threats to carnivorous plants in the wild?
____
____

4. How many carnivorous plant species have been documented? __________

5. List five types of carnivorous plant traps: ______________________
____
____
____
____

Carnivorous Plant Group	Examples:
Pitcher Plants	
Fly Traps	
Sundews	
Bladderworts	
Butterworts	

DESIGN A CARNIVOROUS PLANT Name_______________________

Instructions: Design a plant that has three adaptations that will enable it to defend itself and survive. One adaption must allow it to eat something other than nutrients from the soil. Label each adaptation appropriately and write at least one sentence describing what each adaptation does on your ultimate carnivorous plant. Use your creativity and make a plant that that will not only survive but thrive! Feel free to use the back for more room.

DESIGN A CARNIVOROUS PLANT RUBRIC

Criteria	1	2	3	4
Adaptations	0 adaptations	1 adaptation	2 adaptations	3 adaptations
Labels and Descriptions	Missing all labels and descriptions	Missing most labels and descriptions	Adaptation labels and descriptions are present	Adaptation labels and descriptions are detailed
Creativity	Design lacks color and details	Design is adequate and colorful	Design is nicely drawn but lacks creativity and a unique personal touch	Design is colorful, unique, creative and shows a personal touch

Points scored on rubric	Grade
12	100
11	95
10	90
9	85
8	80
7	75
6	70
5 or below	65

EIGHTH GRADE LESSON PLAN
Crop Rotation

SUBJECT / GRADE LEVEL: 8th Grade - Science.

ACADEMIC STANDARDS AND PERFORMANCE INDICATORS:

Science:

8.E.5 The student will demonstrate an understanding of the processes that alter the structure of Earth and provide resources for life on the planet.

8.E.5A.1 Develop and use models to explain how the process of weathering, erosion, and deposition change surface features in the environment.

LESSON OBJECTIVES

TSW analyze the benefits of crop rotation. TSW evaluate the erosion control that can be provided through proper crop rotation.

MATERIALS:

- *Crop Rotation* student worksheet (attached)
- *Crop Rotation* teacher guide (attached)
- *Crop Rotation Plan* (attached)

ENGAGEMENT:

- **What is crop rotation?**

EXPLORATION:

- The teacher will begin by explaining that planting the same plant in the same area season after season can have negative effects on plant and soil health.

EXPLANATION:

- The teacher will show the video describing crop rotation *Crop Rotation Made Simple–Rotate Your Vegetable Beds for Healthier Produce* by GrowVeg: https://www.youtube.com/watch?v=XeNA6XdMoF8.

ELABORATION:

- If the teacher wishes to extend the lesson, the class can implement the crop rotation schedule created in this lesson.

EVALUATION:

- Student will turn in *Crop Rotation* student worksheet and plan.

RESOURCES:

1. *Crop Rotation Made Simple–Rotate Your Vegetable Beds for Healthier Produce* by GrowVeg: https://www.youtube.com/watch?v=XeNA6XdMoF8.
2. *Crop Rotation* student worksheet (attached).
3. *Crop Rotation* teacher guide (attached).
4. *Crop Rotation* plan (attached).

TEACHER GUIDE
CROP ROTATION

BACKGROUND:

Students should understand that planting the same plant in the same area season after season can have negative effects on plant and soil health.

ENGAGEMENT:

- Teacher will pose the question, "**What is crop rotation?**"

EXPLANATION:

- Guide students through the completion of the *Crop Rotation* student worksheet using the descriptions below:

CROP ROTATION BASICS:

Crop rotation involves not only rotating crops from season to season but rotating entire plant families. For example, radish and carrot are both root vegetables. Pepper and tomato and both fruit vegetables. Beans and peas are in the legume family. Finally, lettuce and cabbage are leaf vegetables. You want to rotate between these four groups each season in the following order: legumes, leaves, roots, and fruits. By rotating crops you can accomplish several goals:

- INSECT CONTROL: Plants in the same family attract the same bugs. Crop rotation is a natural form of pesticide.
- NUTRIENT ADDITION: Plants in the same family have similar eating habits. Crops in the same family deplete the same nutrients from the soil. Therefore, crop rotation allows these nutrients to be replenished in the soil.
- SOIL BUILDING: Legumes actually put nitrogen back into the soil. For this reason we call them soil builders.
- EROSION CONTROL: Crop rotation provides control of wind and water erosion.

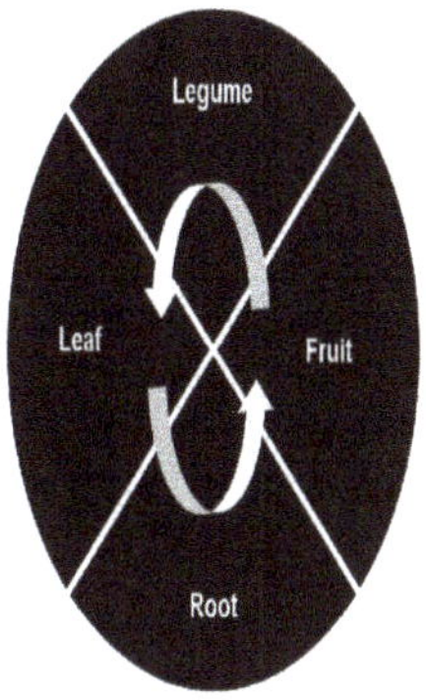

CROP ROTATION

Name__________________________________

Part I: Complete the questions and chart below.

1) What is crop rotation? ____

____

____

2) Which four groups should you rotate between?__________________________

____

____

Benefits of Crop Rotation

Benefit	*How does crop rotation provide this benefit?*
Insect Control	
Nutrient Addition	
Soil Building	
Erosion Control	

CROP ROTATION PLAN

Name________________________________

Instructions: Create a crop rotation plan for your garden using the guide below. For each type of crop, you will fill in four different planting options. The numbers one through four represent growing seasons. You will now be able to follow this growing chart for four seasons of crop rotations for healthy plants and soil.

1) ______________
2) ______________
3) ______________
4) ______________

1) ______________
2) ______________
3) ______________
4) ______________

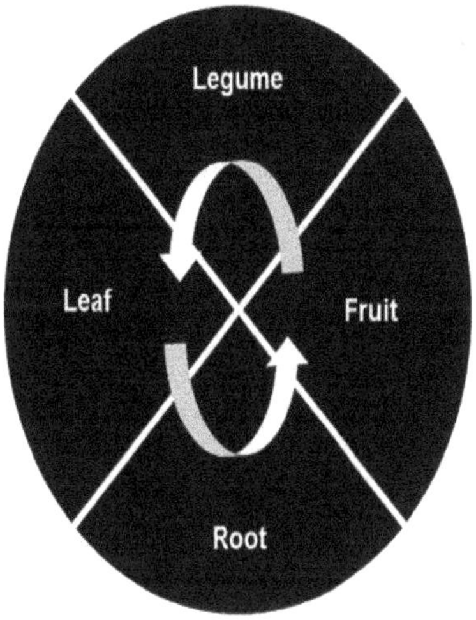

1) ______________
2) ______________
3) ______________
4) ______________

1) ______________
2) ______________
3) ______________
4) ______________

EIGHTH GRADE LESSON PLAN
Fascinating Fossils

SUBJECT / GRADE LEVEL: 8th Grade - Science

ACADEMIC STANDARDS AND PERFORMANCE INDICATORS:

Science:

8.E.6 The student will demonstrate an understanding of Earth's geologic history and its diversity of life over time.

8.E.6A.1 Develop and use models to organize Earth's history including era, period, and epoch according to the geologic time scale using evidence from rock layers.

LESSON OBJECTIVES:

TSW examine plant life that has been fossilized in layers of rocks for millions of years. TSW analyze the differences in prehistoric and modern day plants. TSW discover that some prehistoric plants still inhabit the Earth today.

MATERIALS:

- *Fascinating Fossils* teacher guide (attached)
- *Fascinating Fossils* student worksheet (attached)
- *Prehistoric Garden* student worksheet (attached)

ENGAGEMENT:

- **How do we know which plants were around when dinosaurs roamed the Earth?**

EXPLORATION:

- The teacher will explain that plant fossils give us the ability to understand that plants have reproductively progressed. Spores evolved into cones, and finally, became our modern day seed plants.

EXPLANATION:

- The teacher will begin by showing students the video *Plants Are Cool, Too Episode 2: Fossilized Forests* by Plants Are Cool Too: https://www.youtube.com/watch?v=YfRXDbtkEi0.
- Teacher will guide students through the completion of the *Fascinating Fossils* student worksheet.

ELABORATION:

- An extension to this lesson would be to create a prehistoric garden on your school campus.

EVALUATION:

- Students will turn in the *Fascinating Fossils* and *Prehistoric Garden* student worksheets.

RESOURCES:

1. *Plants Are Cool, Too Episode 2: Fossilized Forests* by Plants Are Cool Too: https://www.youtube.com/watch?v=YfRXDbtkEi0.
2. General Resource: http://www.kidsgardening.org/.
3. *Fascinating Fossils* teacher guide (attached).
4. *Fascinating Fossils* student worksheet (attached).
5. *Prehistoric Garden* student worksheet (attached).

TEACHER GUIDE
FASCINATING FOSSILS

BACKGROUND:

Students should understand that plants fossils help us understand the evolution of plants. Plants have reproductively progressed from spores to cones, and finally, to our modern day seed plants.

ENGAGEMENT:

- Teacher will pose the question, "**How do we know which plants were around when dinosaurs roamed the Earth?**"

EXPLANATION:

- Guide students through the completion of the *Fascinating Fossils* student worksheet by directing students to the *National Gardening Association's* website: http://garden.org/.

FASCINATING FOSSILS STUDENT WORKSHEET KEY:

1. *How do plant fossils form?* Plant fossils are formed when plant matter (such as stems, leaves, roots, spores, seeds, or fruits) is protected from rapid decomposition by being covered with sediment such as clay, mud, sand, and volcanic ash.
2. *Describe four types of plant fossils:*
   a. contain no actual plant remains, but the surrounding sediment retained an imprint of the plant material.
   b. some are preserved as a sculptural likeness of plant material through the process of petrification such as petrified wood.
   c. scientists also obtain information about prehistoric plants from dinosaur excrement.
   d. some contain plant matter that has not fully decomposed.
3. *What reproductive structures were used by the first land plants such as horsetails and ferns?* Spores.
4. *What were the next major group of plants to evolve?* Gymnosperms.
5. *What reproductive structures did they use?* Cones.
6. *What are Angiosperms?* Plants with true flowers.
7. *Give an example of a prehistoric deciduous Gymnosperm:* Bald Cypress.
8. *What prehistoric tree was thought to be extinct but discovered growing in China during World War II?* Dawn Redwood.
9. *Which prehistoric Gymnosperms resemble palms?* Cycads.
10. *Which attractive and fragrant tree is one of the earliest known Angiosperms?* Magnolia.

FASCINATING FOSSILS Name________________________________

Instructions: Answer the following questions.

1. How do plant fossils form?

____

____

____

2. Describe four types of plant fossils:
   a) ____
   b) ____
   c) ____
   d) ____

3. What reproductive structures were used by the first land plants such as horsetails and ferns? ____

4. What were the next major group of plants to evolve? _______________

5. What reproductive structures did they use? ______________________

6. What are Angiosperms? ____

7. Give an example of a prehistoric deciduous Gymnosperm. ___________

____

8. What prehistoric tree was thought to be extinct but discovered growing in China during World War II? ____________________________

9. Which prehistoric Gymnosperms resemble palms? ________________

10) Which attractive and fragrant tree is one of the earliest known Angiosperms? ____

Prehistoric Garden

Name______________________________

Instructions: Complete the following chart with prehistoric plants that still exist today. An extension to this assignment could be the addition of a Prehistoric Garden on your school grounds.

Common Name	Botanical Name	Mature Height

EIGHTH GRADE LESSON PLAN
Garden Geometry

SUBJECT / GRADE LEVEL: 8th Grade - Science.

ACADEMIC STANDARDS AND PERFORMANCE INDICATORS:

Science:

8-1.1 Generate and solve complex, abstract problems that involve modeling physical, social, or mathematical phenomena.

8-1.3 Use inductive and deductive reasoning to formulate mathematical arguments.

8-1.4 Understand equivalent symbolic expressions as distinct symbolic forms that represent the same relationship.

8-1. Generalize connections among a variety of representational forms and real world situations.

LESSON OBJECTIVES:

TSW analyze the plants within the garden that naturally exhibit a Fibonacci sequence within their petals, leaves, and seeds.

MATERIALS:

- *Garden Geometry* student worksheet (attached)
- *Garden Geometry* teacher guide (attached)

ENGAGEMENT:

- **What is a Fibonacci sequence?**

EXPLORATION:

- Begin by explaining to students that plants can grow cells in spirals by turning after producing each new cell. Otherwise, the cells would form in a straight line.

EXPLANATION:

- The teacher will begin by showing students the video *Doodling in Math: Spirals, Fibonacci, and Being a Plant* [1 of 3] by Vi Hart: https://www.youtube.com/watch?v=ahXIMUkSXX0.

- The teacher will guide students through the completion of the *Garden Geometry* student worksheet.

ELABORATION:

- An extension to this lesson would be collecting items from the garden that contain Fibonacci numbers. This would include flowers, leaves, and some seed pods, such as pine cones.

EVALUATION:

- Students will turn in the *Garden Geometry* student worksheet.

RESOURCES:

1. *Doodling in Math: Spirals, Fibonacci, and Being a Plant* [1 of 3] by Vi Hart: https://www.youtube.com/watch?v=ahXIMUkSXX0.
2. General Resource: https://www.Mathsisfun.com/numbers/nature-golden-ratio-fibonacci.html.
3. *Garden Geometry* teacher guide (attached).
4. *Garden Geometry* student worksheet (attached).

TEACHER GUIDE
GARDEN GEOMETRY

BACKGROUND:

Students should understand that nature by design has many things that contain both Fibonacci numbers and the Golden ratio.

ENGAGEMENT:

- Teacher will pose the question, "**What is a Fibonnaci sequence?**"

EXPLANATION:

- Guide students through the completion of the *Garden Geometry* student worksheet by directing students to the following website: http://www.mathsisfun.com/numbers/nature-golden-ratio-fibonacci.html.
- Begin by explaining to students that plants can grow cells in spirals by turning after producing each new cell. Otherwise, the cells would form in a straight line.
- There is a calculator on this website that models spiral plant formation. Plug in a few numbers to show students how nature uses the Golden Ratio.

***GARDEN GEOMETRY* KEY:**

1) *What pattern will be created by most numbers that are simple fractions?* A pattern of lines stacking up which makes gaps.

2) *How do you design a spiral without gaps?* By using the Golden Ratio, 1.61803.

3) *What type of number is the Golden Ratio?* An irrational number.

4) *What is a Fibonacci Number?* Each number is the sum of the two numbers before it.

5) *When we take any two successive (one after the other) Fibonacci Numbers, what is their ratio near?* It is always very close to the Golden Ratio.

6) *In addition to sunflower seeds, what other plant parts exhibit Fibonacci Numbers and the Golden Ratio?* Leaves, branches and petals.

7) *What is one theory as to why plants grow in these spirals?* So that new leaves don't block the sun from older leaves, or so that the maximum amount of rain or dew gets directed down to the roots.

GARDEN GEOMETRY Name________________________________

Instructions: Answer the following questions.

1) What pattern will be created by most numbers that are simple fractions?

    ____

2) How do you design a spiral without gaps?

    ____

3) What type of number is the Golden Ratio?

    ____

4) What is a Fibonacci number?

    ____

5) When we take any two successive *(one after the other)* Fibonacci Numbers, what is their ratio near?

    ____

6) In addition to sunflower seeds, what other plant parts exhibit Fibonacci numbers and the Golden Ratio?

    ____

7) What is one theory as to why plants grow in these spirals?

    ____

    ____

    ____

    ____

EIGHTH GRADE LESSON PLAN

Light Waves

SUBJECT / GRADE LEVEL: 8th Grade - Science.

ACADEMIC STANDARDS AND PERFORMANCE INDICATORS:

Science:

8.P.3 The student will demonstrate an understanding of the properties and behaviors of waves.

8.P.3A.1 Construct explanations of the relationship between matter and energy based on the characteristics of mechanical and light waves.

LESSON OBJECTIVES:

TSW investigate the properties of light waves and their importance to plant growth. TSW analyze the way plants use different light spectrum in different ways.

MATERIALS:

- *Light Waves* student worksheet (attached)
- *Light Waves* teacher guide (attached)

ENGAGEMENT:

- **What color is light?**

EXPLORATION:

- The teacher will begin by explaining that sunlight contains many colors but plants use the different light spectrum in different ways.

EXPLANATION:

- The teacher will follow up the engagement question with the video titled *Light Reactions* by Biology Videos: https://www.youtube.com/watch?v=D6tbOBhbOZU.

ELABORATION:

- If the teacher wishes to extend the lesson, the class can conduct the experiment described on the following website:
http://www.udel.edu/MERL/Outreach/Teacher 2–s 20Guide/2. 20Effects 20of 20color 20on 20plant 20growth 20TE.pdf.

EVALUATION:

- Student will turn in the *Light Waves* student worksheet.

RESOURCES:

1. *Light Reactions* by Biology Videos:
https://www.youtube.com/watch?v=D6tbOBhbOZU.
2. General Resource:
http://homeguides.sfgate.com/plant-light-spectrum-growing-flowering-plants-72801.html.
3. General Resource:
http://www.udel.edu/MERL/Outreach/Teacher 2–s 20Guide/2. 20Effects 20of 20color 20on 20plant 20growth 20TE.pdf.
4. *Light Waves* teacher guide (attached).
5. *Light Waves* student worksheet (attached).

TEACHER GUIDE
LIGHT WAVES

BACKGROUND:

Students should understand that plants absorb different wavelengths of light for different types of plant growth.

ENGAGEMENT:

- Teacher will pose the question, "**What color is light?**"

EXPLANATION:

- Guide students through the completion of the *Light Waves* student worksheet by directing students to the following website: http://homeguides.sfgate.com/plant-light-spectrum-growing-flowering-plants-72801.html.

LIGHT WAVES STUDENT WORKSHEET KEY:

1. *Which colors are included in the visible light spectrum*? Red, orange, yellow, green, blue, indigo and violet.
2. *Which colors of the light spectrum play the biggest role in plant growth and flowering?* Colors on either end of the spectrum play the greatest role in plant growth and flowering.
3. *Why do leaves appear to be green in color?* Wavelengths in the red and blue spectrum are absorbed in greater amounts while more green and yellow light is reflected, giving the leaves their characteristic green color.
4. *What is cool light?* It is blue wavelengths from 430 to 450 nm.
5. *What type of plant growth is encouraged through cool light?* These wavelengths encourage vegetative growth through strong root growth and intense photosynthesis.
6. *What is warm light?* Red wavelengths are known as warm light and they are naturally more prevalent in sunlight during the shorter days of fall and winter.
7. *What type of plant growth is encouraged by the red wavelengths of light?* These wavelengths encourage stem growth, flowering and fruit production, and chlorophyll production.
8. *Do plants absorb any light from the green or yellow wavelengths of the spectrum?* While most of the absorbed wavelengths are in the red and blue ranges, recent research shows that plants do absorb some green and yellow light and use it in the photosynthesis process. A light source that provides light in the entire visible range will better meet the needs of the plant.

LIGHT WAVES Name______________________________

Instructions: Answer the following questions.

1) Which colors are included in the visible light spectrum?

____

2) Which colors of the light spectrum play the biggest role in plant growth and flowering?

____

3) Why do leaves appear to be green in color?

____

____

4) What is cool light?

____

5) What type of plant growth is encouraged through cool light?

____

6) What is warm light?

____

____

7) What type of plant growth is encouraged by the red wavelengths of light?

____

8) Do plants absorb any light from the green or yellow wavelengths of the spectrum?

____

____

EIGHTH GRADE LESSON PLAN
Plants Can Tell Time

SUBJECT / GRADE LEVEL: 8th Grade - Science.

ACADEMIC STANDARDS AND PERFORMANCE INDICATORS:

Science:

8.E.4 The student will demonstrate an understanding of the universe and the predictable patterns caused by Earth's movement in the solar system.

8.E.4B.3 Develop and use models to explain how seasons, caused by the tilt of Earth's axis as it orbits the sun, affects the length of the day and the amount of heating on Earth's surface.

LESSON OBJECTIVES:

TSW evaluate the benefits to humans and agriculture created by the Earth's tilt on its axis. TSW examine the mechanisms that allow food crops to thrive based on day length and seasons.

MATERIALS:

- *Plants Can Tell Time* student worksheet (attached)
- *Plants Can Tell Time* teacher guide (attached)
- *Seasonal Crop Chart* worksheet (attached)

ENGAGEMENT:

- **How do plants in your garden tell time?**

EXPLORATION:

- The teacher will explain that if the Earth was not tilted on its axis, we would have twelve hours of daylight and twelve hours of night throughout the year everywhere. However, because the Earth is tilted on its axis, we have our four seasons which affect the food crops that we grow.

EXPLANATION:

- The teacher will show *How Plants Tell Time* by Discovery News: https://www.youtube.com/watch?v=klq1Gmdbh-g.
- The teacher will guide students through the completion of the *Plants Can Tell Time* student worksheet and the *Seasonal Crop Chart* worksheet.

ELABORATION:

- If the teacher wishes to extend the lesson, the class can plant the spring and fall plantings based on the *Seasonal Crop Chart* worksheet.

EVALUATION:

- Students will turn in the *Plants Can Tell Time* student worksheet and the *Seasonal Crop Chart* worksheet.

RESOURCES:

1. *How Plants Tell Time* by Discovery News: https://www.youtube.com/watch?v=klq1Gmdbh-g.
2. General Resource: http://garden.org/.
3. General Resource: http://www.livescience.com/18972-earth-seasons-tilt.html.
4. General Resource: https://www.clemson.edu/extension/hgic/plants/vegetables/gardening/hgic1256.html.
5. *Plants Can Tell Time* teacher guide (attached).
6. *Plants Can Tell Time* student worksheet (attached).
7. *Seasonal Crop Chart* worksheet (attached).

TEACHER GUIDE
PLANTS CAN TELL TIME

BACKGROUND:

Students should understand that if the Earth was not tilted on its axis then we would have twelve hours of daylight and twelve hours of night throughout the year everywhere. However, because the Earth is tilted on its axis we have our four seasons which affect the food crops that we grow.

ENGAGEMENT:

- Teacher will pose the question, "**How do plants in your garden tell time?**"

EXPLANATION:

- Guide students through the completion of the *Plants Can Tell Time* student worksheet Part I by directing students to the *National Gardening Association's* website: http://garden.org/.

PLANTS CAN TELL TIME STUDENT WORKSHEET KEY:

The Photoperiodic Response

1. *How can plants tell time?* Plants can measure and respond to the relative lengths of day and night. So they must have some capacity to "count" numbers of hours. Scientists have determined that a plant's leaves are responsible for doing the counting.
2. *What protein pigment do leaves have that make them sensitive to light?* Phytochrome.
3. *How does phytochrome allow plants to detect when it is day and night?* This pigment occurs in two stable forms, one sensitive to visible red light, one sensitive to far-red light (on the edge of the visible spectrum). A pigment molecule can convert from one form to the other, depending on the type of light it receives. In total darkness, however, the far-red-sensitive form slowly reverts to the other form. The length of total darkness, then, determines the ratio of the two forms. This is how plants "count" the number of hours of darkness.
4. *What happens when a plants darkness requirement is met?* The leaves release certain plant growth regulators. The substances travel from the leaves through the stem to the apical buds, stimulating some of those buds to switch from leaf production to flower production.

- Next, guide students through the completion of the *Plants Can Tell Time* student worksheet Part II by directing students to the following website: http://www.livescience.com/18972-earth-seasons-tilt.html.

WHAT WOULD HAPPEN WITHOUT THE EARTH'S TILT KEY:

1. *Where would most humans live?* Near the tropical midsection of the Earth.
2. *Why would erosion be a problem?* There would be too much rain which would erode soil in cleared areas for farming and the nutrients would leach below the root level making the land infertile for crops.
3. *Why would insects and diseases be a problem for both humans and food crops?* Disease pathogens thrive in warm, humid environments. Winter protects the world from many insects as well.
4. *What type of Winter does wheat need to grow?* Cool or cold.
5. *What are some other essential food crops that prefer a cool winter?* Corn, potatoes, oats, and barley.

EVALUATION:

- Guide students through the completion of *Seasonal Crops* worksheet using the following website for cool and warm season crop suggestions: http://www.clemson.edu/extension/hgic/plants/vegetables/gardening/hgic1256.html.

Plants Can Tell Time Name________________________________

The Photoperiodic Response

1. How can plants tell time?

____

____

2. What protein pigment do leaves have that make them sensitive to light?

____

3. How does phytochrome allow plants to detect when it is day and night?

____

____

4. What happens when a plants darkness requirement is met?

____

____

What Would Happen Without the Earth's Tilt

1. Where would most humans live?

____

____

2. Why would erosion be a problem?

____

____

3. Why would insects and diseases be a problem for both humans and food crops?

____

____

4. What type of Winter does wheat need to grow?

____

5. What are some other essential food crops that prefer a cool winter?

____

Seasonal Crop Chart

Name________________________________

Instructions: Create a seasonal planting chart based on the Spring and Fall planting dates as well as your location.

Spring Crops

Crop	Days to Harvest	Planting Depth

Fall Crops

Crop	Days to Harvest	Planting Depth

EIGHTH GRADE LESSON PLAN
Soil Profile

SUBJECT / GRADE LEVEL: 8th Grade - Science.

ACADEMIC STANDARDS AND PERFORMANCE INDICATORS:

Science:

8.E.5 The student will demonstrate an understanding of the processes that alter the structure of Earth and provide resources for life on the planet.

8.E.5A.1 Develop and use models to explain how the process of weathering, erosion, and deposition change surface features in the environment.

LESSON OBJECTIVES:

TSW analyze the horizons of a soil profile to determine the differences between the layers created by weathering, erosion and deposition.

MATERIALS:

- *Soil Profile* student worksheet (attached)
- *Soil Profile* teacher guide (attached)
- *Soil Profile Drawing and Rubric* (attached)

ENGAGEMENT:

- **What is a profile?**

EXPLORATION:

- The teacher will begin by explaining that weathering, erosion and deposition can alter the layers within the soil.

EXPLANATION:

- The teacher will guide students through the completion of the *Soil Profile* student worksheet and drawing.

ELABORATION:

- If the teacher wishes to extend the lesson, the class can excavate an area of native soil to a depth of three to five feet to analyze the horizons within the soil profile.

EVALUATION:

- Students will turn in the *Soil Profile* student worksheet and drawing.

RESOURCES:

1. Plant and Soil Sciences eLibrary: http://passel.unl.edu/pages/.
2. *Soil Profile* teacher guide (attached).
3. *Soil Profile* student worksheet (attached).
4. *Soil Profile Drawing and Rubric* (attached).

TEACHER GUIDE
SOIL PROFILE

BACKGROUND:

Students should understand that weathering, erosion and deposition can alter the layers within the soil.

ENGAGEMENT:

- Teacher will pose the question, "**What is a profile?**"

EXPLANATION:

- Guide students through the completion of the *Soil Profile* student worksheet by directing students to the following website: http://passel.unl.edu/pages/informationmodule.php?idinformationmodule=1130447025&topicorder=4

SOIL PROFILE WORKSHEET KEY:

1. *What are soil horizons?* Through the interactions of soil building processes, the components of soil are reorganized into visibly, chemically, and/or physically distinct layers, referred to as horizons.

2. *What are the horizons contained within the soil profile?* A,B,C,O,E,R

Horizon	Name	Description
A	Topsoil	Rich in organic matter, darker in color.
B	Subsoil	Clay, soluble salts, and iron.
C	Parent Material	Little or no alterations of parent material.
O	Organic Matter	At least 20% Organic Matter, black or dark brown.
E	Horizon of Losses	Lighter in color.
R	Bedrock	Underlying bedrock.

SOIL PROFILE Name ____________________

Instructions: Answer the following questions and complete the chart below.

1) What are soil horizons?

____

____

____

____

2) What are the horizons contained within the soil profile?

____

Horizon	Name	Description

SOIL PROFILE DRAWING RUBRIC

Criteria	1	2	3	4
Horizons	3 or less horizons	4 horizons	5 horizons	6 horizons
Label and Description	Missing most labels and descriptions	Some labels and some descriptio-ns are correct	All labels and most descriptio ns are correct.	All labels and descriptio-ns are correct.
Creativity	Design lacks color and details	Design is adequate and colorful	Design is nicely drawn but lacks creativity and a unique personal touch	Design is colorful, unique, creative and shows a personal touch

Points scored on rubric	Grade
12	100
11	95
10	90
9	85
8	80
7	75
6	70
5 or below	65

SOIL PROFILE DRAWING Name ______________________________

Instructions: Draw and label a soil profile in the area below. Include horizons O, A, B, C, E and R. Your drawing should show a distinct difference between the horizons within the profile.

EIGHTH GRADE LESSON PLAN
Terrific Textures

SUBJECT / GRADE LEVEL: 8th Grade - Science - Duration: Two 60 minute lessons.

ACADEMIC STANDARDS AND PERFORMANCE INDICATORS:

Science:

8.E.5 The student will demonstrate an understanding of the processes that alter the structure of Earth and provide resources for life on the planet.

8.E.5A.1 Develop and use models to explain how the process of weathering, erosion, and deposition change surface features in the environment.

LESSON OBJECTIVES:

TSW analyze soil samples to determine the soil texture based on the amount of sand, silt, or clay. TSW evaluate the effects of different types of soil texture on the characteristics of an ecosystem using evidence from soil samples.

MATERIALS:

- Soil
- Newspaper
- Spoon or hand trowel
- Tall jar with lid
- Spray bottle of water
- *Soil Texture Jar Lab* student worksheet (attached)
- *Determining Soil Texture* student worksheet (attached)
- *Soil Texture Hands On Lab* student worksheet (attached)

ENGAGEMENT:

- **What's in a Soil?**
- Students will draw a pie chart hypothesizing the components of soil. *Correct Pie Chart: 25 Air, 25 Water, 45 Sand, Silt and Clay, 5 Organic Matter.*

EXPLORATION:

- DAY ONE: The teacher will begin by having students draw a pie chart representing the components of soil.
- DAY TWO: Students will work in pairs to determine soil texture using the ribbon method.

EXPLANATION:

- DAY ONE: The teacher will display the soil texture triangle and explain that the twelve different soil textures are determined by the percentages of sand, silt and clay in a soil sample.
- Students will begin *Soil Texture Jar Lab* observing that soil particles settle according to their size, with sand being the largest and clay being the smallest.
- The teacher will direct students to the following website in order to complete the *Determining Soil Texture* student worksheet: http://www.nrcs.usda.gov/wps/portal/nrcs/detail/soils/survey/?cid=nrcs142p2_05416–.
- DAY TWO: The teacher will show video describing the ribbon test of determining soil texture titled *Kansas State University Extension Eye on Agriculture Today: Soil Texture* By Feel Video: https://www.youtube.com/watch?v=IOyaBxj–6–s.
- Each pair of students will complete the soil texture lab by collecting a sample in their hands and performing the soil ribbon test in their hands. As a class, students will compare the results of their ribbon tests.

ELABORATION:

- If the teacher wishes to extend the lesson, the class can perform an experiment in which students plant carrots or radishes in containers with a soil containing a high percentage of sand, as well as a second container with a soil containing a high percentage of clay.
- Students should make predictions about the size and shape of mature root vegetables grown in sand versus clay.
- Once harvested, students should record their observations against their original predictions.

EVALUATION:

- DAY ONE: Students will turn in the *Determining Soil Texture* student worksheet.
- DAY TWO: Students will turn in their results from the *Soil Texture Jar Lab* and the *Soil Texture Hands On Lab.*

RESOURCES:

1. *Kansas State University Extension Eye on Agriculture Today: Soil Texture* by Feel Video: https://www.youtube.com/watch?v=IOyaBxj–6–s.
2. General Resource: http://www.nrcs.usda.gov/wps/portal/nrcs/detail/soils/survey/?cid=nrcs142p2_05416–.
3. *Soil Texture Jar Lab* student worksheet (attached).
4. *Determining Soil Texture* student worksheet (attached).
5. *Soil Texture Hands On Lab* student worksheet (attached).

Determining Soil Texture:

Using the soil texture triangle, find your percentages of sand, silt and clay to determine the texture of your sample. (*See example at bottom of page*)

According to the Soil Texture Triangle, What type of soil texture do you have?

____

	% sand	% silt	% clay	Soil Texture
Example:	60	10	30	Sandy Clay Loam

You can find the soil texture triangle on the following website:
http://www.nrcs.usda.gov/wps/portal/nrcs/detail/soils/survey/?cid=nrcs142p2_054167

Using the soil texture triangle, find your percentages of sand, silt and clay to complete the chart below.

Percentage of Silt	Percentage of Sand	Percentage of Clay	Soil Texture
10	60	30	
20	22	58	
56	10	34	
47	6	47	
6	52	42	
70	17	13	
32	58	10	

Soil Texture Hands- On Lab

Instructions: *Students should work in pairs. One student should perform the ribbon test while the other student reads and interprets the soil texture key.*

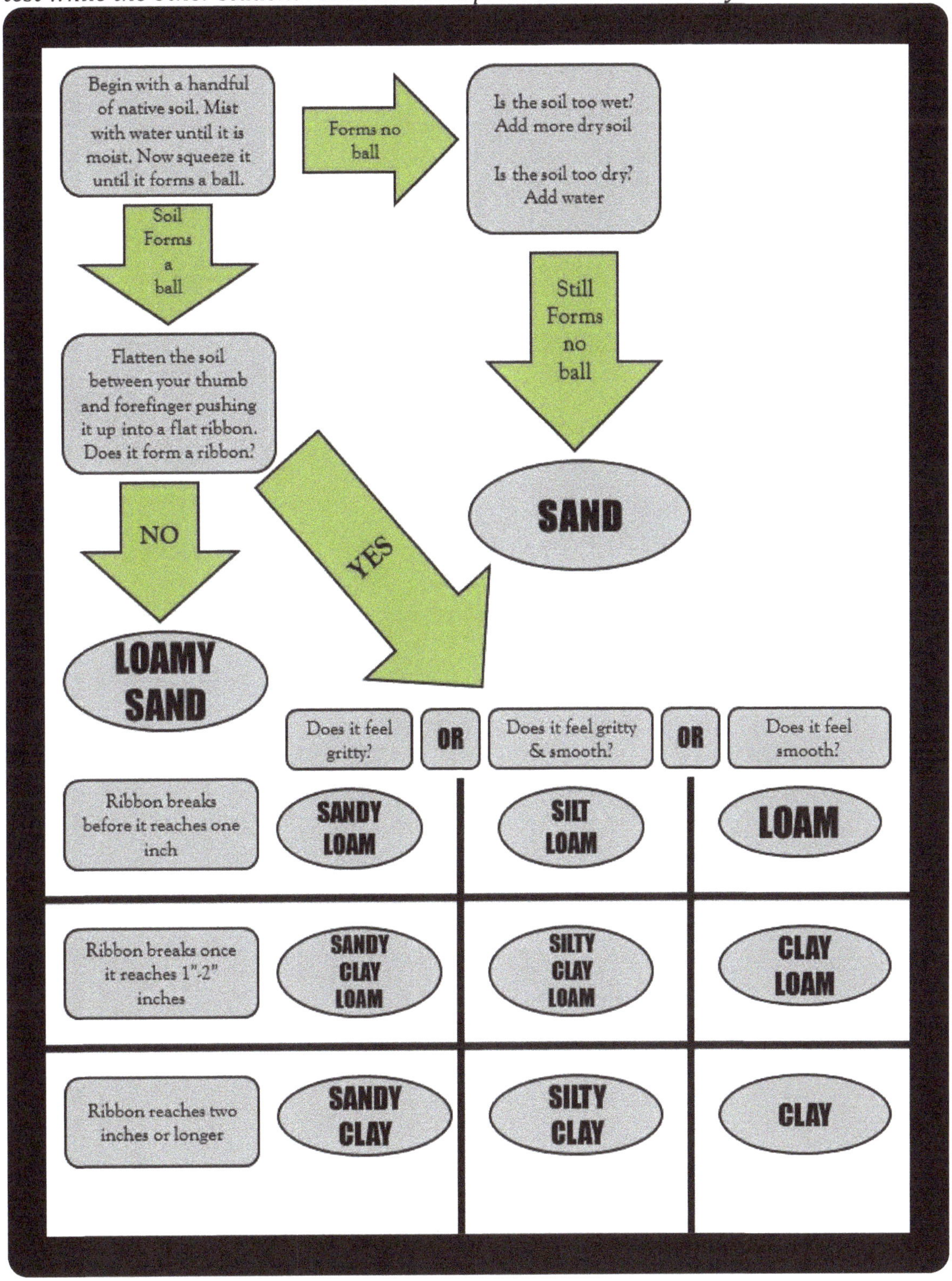

1. Go outside and collect a native soil sample from the ground. Spread out soil on newspaper in the sun to dry. Remove all debris including roots and rocks.
2. Take a hand trowel or spoon and crush the soil into fine particles.
3. Fill a tall glass jar one-third full of soil.
4. Add water until jar is two inches from the top of the jar.
5. Add a teaspoon of liquid dish detergent.
6. Put on lid and have each student shake the jar vigorously for one minute (*10-20 minutes total)*
7. Set jar on level surface and observe the soil particles settling based on their size. (*Notice that the particles will settle with the largest one, sand, on the bottom followed by silt then clay on top*)
8. After two minutes, make a mark on the jar indicating the depth of the layer o sand.
9. At the end of the lesson, make a second mark indicating the depth of the laye of silt. (*Must wait no less than thirty minutes*)
10. Begin Day Two of this lab by marking the depth of the clay layer in the jar (*Depending on your soil type it could take more than one day for your water to become clear again*)
11. Once water has become clear and the three layers are distinguishable, measure the thickness of each layer.

ulations:

th of Clay Layer (*top*): ____________________

th of Sand Layer (*bottom*): ____________________

th of Silt Layer (*middle*): ____________________

sure the TOTAL depth of all three layers: __________________________

Layer (*Clay*)/Total Depth = __________ % of Clay

dle Layer (*Silt*)/Total Depth = __________ % of Silt

om Layer (*Sand)*/Total Depth = __________ % of Sand

EIGHTH GRADE LESSON PLAN
Wide World of Worms

SUBJECT / GRADE LEVEL: 8th Grade - Science.

ACADEMIC STANDARDS AND PERFORMANCE INDICATORS:

Science:

8.E.5 The student will demonstrate an understanding of the processes that alter the structure of Earth and provide resources for life on the planet.

8.E.5A.1 Develop and use models to explain how the process of weathering, erosion, and deposition change surface features in the environment.

LESSON OBJECTIVES:

TSW understand that vermicomposting creates organic matter which can be added to soil in order to improve soil structure as well as aid in the reduction of erosion.

MATERIALS:

- *Wide World of Worms* student worksheet (attached)
- *Wide World of Worms* teacher guide (attached)
- *Design A Vermicompost Bin* worksheet (attached)

ENGAGEMENT:

- **What is vermicomposting?**

EXPLORATION:

- The teacher will explain the role of compost in erosion control and plant health.

EXPLANATION:

- The teacher will begin the lesson by showing the video *Vermicomposting: Using Worms to Make Compost for Your Garden* by UMDHGIC: https://www.youtube.com/watch?v=1CTOvPlc8xk.

- Guide students through the completion of the *Wide World of Worms* and *Design A Vermicompost Bin* student worksheets.

ELABORATION:

- An extension to this lesson would be building a vermicomposting bin for the school garden.

EVALUATION:

- Students will turn in *Wide World of Worms* student worksheet and *Design A Vermicompost Bin* worksheet.

RESOURCES:

1. *Vermicomposting: Using Worms to Make Compost for Your Garden* by UMDHGIC: https://www.youtube.com/watch?v=1CTOvPlc8xk.
2. Composting Resource: http://www.wormpoop.com/composting/composting.htm.
3. *Wide World of Worms* teacher guide (attached).
4. *Wide World of Worms* student worksheet (attached).
5. *Design a Vermicompost Bin* worksheet (attached).

TEACHER GUIDE
WIDE WORLD OF WORMS

BACKGROUND:

Students should understand the erosion control and soil health are important components to the successful growing of food.

ENGAGEMENT:

- Teacher will pose the question, "**What is vermicomposting?**"

EXPLANATION:

- Guide students through the completion of the *Wide World of Worms* student worksheet by directing students to the following website: http://www.wormpoop.com/composting/composting.htm

VERMICOMPOST WORKSHEET KEY:

1. *What is Vermiculture?* Is the process of using worms to decompose organic food waste, turning the waste into a nutrient-rich material capable of supplying necessary nutrients to help sustain plant growth.
2. *What are the benefits of creating vermicompost?* It saves water, energy, landfills, and helps rebuild the soil.
3. *How does Vermicompost benefit the soil?* Vermicompost improves soil structure, texture, and aeration as well as increasing its water-holding capacity. Your plants will grow stronger and have deeper root systems for better drought tolerance and disease resistance. Adding compost to soil also aids in erosion control, promotes soil fertility, and stimulates healthy root development in plants.
4. *What do you feed worms?* Fruit and vegetable scraps, grass clippings, leaves and anything but meat and dairy products.
5. *What type of worms are best for Vermicomposting?* Red worms.
6. *What type of care do red worms need?* Food, moisture, oxygen and a dark place to live.

WIDE WORLD OF WORMS Name ______________________

Instructions: Answering the following questions.

1) What is Vermiculture?

____

____

____

2) What are the benefits of creating vermicompost?

____

____

3) How does Vermicompost benefit the soil?

____

____

____

4) What do you feed worms?

____

____

____

5) What type of worms are best for Vermicomposting?

____

6) What type of care do red worms need?

____

____

DESIGN A VERMICOMPOST BIN Name ______________________________

Instructions: In the space below, draw the design for a school vermicomposting bin. Include your list of materials as well as assemble instructions.

Assembly Instructions and Materials List:

____

____

____

____

____

EIGHTH GRADE LESSON PLAN
Scavenger Hunt

SUBJECT / GRADE LEVEL: 8th Grade - Science.

ACADEMIC STANDARDS AND PERFORMANCE INDICATORS:

Science:

8.E.5 The student will demonstrate an understanding of the processes that alter the structure of Earth and provide resources for life on the planet.

8.E.5A.1 Develop and use models to explain how the process of weathering, erosion, and deposition change surface features in the environment.

8.E.6 The student will demonstrate an understanding of Earth's geologic history and its diversity of life over time.

8.E.6A.1 Develop and use models to organize Earth's history including era, period, and epoch according to the geologic time scale using evidence from rock layers.

8.P.3 The student will demonstrate an understanding of the properties and behaviors of waves.

8.P.3A.1 Construct explanations of the relationship between matter and energy based on the characteristics of mechanical and light waves.

Math:

8-1.1 Generate and solve complex, abstract problems that involve modeling physical, social, or mathematical phenomena.

8-1.3 Use inductive and deductive reasoning to formulate mathematical arguments.

8-1.4 Understand equivalent symbolic expressions as distinct symbolic forms that represent the same relationship.

8-1.7 Generalize connections among a variety of representational forms and real world situations.

LESSON OBJECTIVES:

TSW analyze the benefits of crop rotation. TSW evaluate the erosion control that can be provided through proper crop rotation. TSW analyze the differences in prehistoric and modern day plants. TSW discover that some prehistoric plants still inhabit the Earth today. TSW analyze the plants within the garden that naturally exhibit a Fibonnaci sequence within their petals, leaves, and seeds. TSW investigate the properties of light waves and their importance to plant growth. TSW analyze the horizons of a soil profile to determine the differences between the layers created by weathering, erosion and deposition. TSW analyze soil samples to determine the soil texture based on the amount of sand, silt, or clay.

MATERIALS:

- *Crop Rotation* student worksheet
- *Crop Rotation Plan*
- *Fascinating Fossils* student worksheet
- *Prehistoric Garden* student worksheet
- *Garden Geometry* student worksheet
- *Light Waves* student worksheet
- *Soil Profile* student worksheet
- *Soil Profile* drawing
- *Determining Soil Texture* Sstudent worksheet
- *Scavenger Hunt* student worksheet attached)
- Paper lunch sacks to collect items

****Students should complete the lessons listed above to gain prior knowledge in order to complete the scavenger hunt.*

ENGAGEMENT:

- **How many plant terms can you name?**

EXPLORATION:

- The teacher will poll students for plant terms until all of the terms for the scavenger hunt have been stated.

EXPLANATION:

- Using the *Scavenger Hunt* chart, the teacher will guide students through the explanation of all terms in the chart.
- After reviewing the *Scavenger Hunt* chart, students will go to the garden in order to find the items.

ELABORATION:

- Students can compost all of the items from the scavenger hunt including the chart and the paper bag. If there is nowhere to compost the items, this could also be turned into a digital scavenger hunt where students simply take pictures of the items in the chart.

EVALUATION:

- Students will show teacher items in the bag as the teacher checks them off of the *Scavenger Hunt* chart.

RESOURCES:

1. *Scavenger Hunt* student worksheet (attached).

SCAVENGER HUNT

Name ______________________________

Instructions: Find all of the items listed below:

Item	Where did you find it?
1. A root vegetable	
2. A legume	
3. A leaf vegetable	
4. One prehistoric plant	
5. A seed pod that exhibits a Fibonacci sequence of numbers	
6. Leaves that exhibit a Fibonacci sequence of numbers	
7. A flower that exhibits a Fibonacci sequence of numbers	
8. A plant that absorbs blue and red wavelengths of light	
9. A sample from the A horizon of the soil profile	
10. Soil Sample with the texture of Sandy Loam or Sandy Clay	

HARVEST TASTING GUIDE

Harvest Tasting Guide

Select cooking method based on time and supplies available.

SOME OF THE COOKING SUPPLIES INCLUDE:

- Seasonal harvest
- Knife
- Cutting board
- Baking sheet
- Large sauté pan
- Medium pot
- Microwave safe dish
- Variety of seasonings
- Oil
- Oven
- Stove
- Microwave.

Rinse all vegetables before preparing. Scrub all root vegetables. Reserve scraps for composting or to make a vegetable stock.

- Vegetables are done when their color is vibrant and they maintain a slight crunch.
- Use cooled or leftover vegetables in salads.
- *Bake*: Line baking sheet with foil. Heat oven to 425°. Bake until tender.
- *Sauté*: Heat a small amount of canola oil in a large sauté pan over medium heat and cook according to directions below.
- *Steam*: Fill a pot with 1½ inches water. Add vegetables and simmer according to directions.
- *Microwave*: Fill a microwave safe dish with ¼ inch water and cook covered. Allow to sit 1 minute before preparing.

		Cooking Method			
Vegetable	Preparation	Bake	Sauté	Steam	Microwave
Asparagus	Cut white end off. Keep whole or cut into 1 inch pieces. Eat raw or cook.	Toss with oil, salt, and pepper. Bake at 425° for 10-15 minutes. Enjoy as is or toss with light vinaigrette.	Heat 1 tablespoon oil over medium heat in large sauté pan. Sauté 4-6 minutes with spices, salt and pepper.	Melt butter in 1 inch of water. Place in steam basket and cook 5-10 minutes.	Cook 3-5 minutes. Toss with butter or oil and spices.
Beans	Shell if needed. Don't over stir when cooking.		Cook in small amount of oil for 5-8 minutes.	Cook 5-8 minutes to maintain color and flavor.	Cook 3-5 minutes.
Broccoli, Cauliflower	Cut florets off stem into even sizes. For an easy salad toss with 2 tablespoons vinegar, 1 cup light mayo, ¼ cup sugar, diced onion, sunflower seeds, and raisins. *Peel the stem, slice and enjoy with dip.	Toss with oil, garlic powder, salt, and pepper. Bake at 425° for 12-15 minutes. *Try smoked paprika for more flavor.	Heat 1 tablespoon oil over medium heat in a large sauté pan. Sauté 5-8 minutes with spices, salt, and pepper. *Start with sautéed onions and garlic.	In steam basket, and cook 5-8 minutes. Broccoli should be bright green.	Cook 5-7 minutes. Toss with butter or oil and spices.
Brussel Sprouts	Cut off stem. Slice In half or in quarters.	Toss with oil, garlic powder, salt, and pepper. Bake at 425° for 25-35 minutes. *Add sliced onions and apples halfway through cooking.	Heat 1 tablespoon oil over medium heat in a large sauté pan. Sauté 5-8 minutes with spices, salt, and pepper. *Add cooked bacon.	Place in steam basket, and cook 5-8 minutes. *Once cool, separate leaves and use leaves in a salad.	Cook 5-7 minutes. Toss with butter or oil and spices.
Cabbage	Cut according to cooking method. Slice into ¼ inch strips. Toss with ½ cup sugar, 1/3 cup vinegar, 1/3 cup oil, and 2 teaspoons salt. For coleslaw: slice into ¼ inch strips and toss with ½ cup sugar.	Cut into 1 inch "steaks." Rub with smashed clove of garlic, oil, salt, and pepper. Roast 1 hour flipping halfway through.	Cut into quarters and remove hard stem. Cube cabbage. Heat oil over medium heat, sauté onions until translucent, add cabbage, salt, and pepper. Cook until soft. Add ¾ cup vegetable stock, 3 tablespoons balsamic vinegar, and ¼ cup craisins. Cook 20 minutes.	Cut into quarters and remove hard stem. Slice into 1 inch wide strips. In a stock pot, melt 2 tablespoons butter, 2 tablespoons oil, ½ teaspoon salt, ½ teaspoon pepper. Add 1 cup chicken broth. Stir in cabbage strips. Simmer 10-12 minutes until tender.	Cook 8 minutes. Add butter and salt.

		Cooking Method			
Carrots	Eat whole raw. Peel and cut into strips or slices. Shred and add to coleslaw or salad. *Use stem in stocks.	Depending on size keep whole or cut into strips. Toss with oil, salt, and pepper. Bake 15-20 minutes flipping halfway through cooking. *Make a mixture of parsnip and carrots for tasty "fries."	Cut into ¼ inch slices. Heat 1 tablespoon oil in large sauté pan. Add carrots and a sprinkle of salt and pepper. Cook 12 minutes or until tender. Add 1 tablespoon chopped, fresh rosemary and 1 ½ tablespoons honey.	Steam sliced or whole small carrots in a small amount of water for 10 minutes. Drain and toss with butter, 2 cloves minced garlic, 1 teaspoon fresh, minced ginger. Sprinkle lime juice on top and chopped fresh cilantro.	Cook for 7 minutes or until tender. Season with butter, herbs, and spices.
Celery	Cut off root end. Cut into sticks. Reserve the leaves for tasting. Eat with dips or nut butter or add to salads.	Bake into vegetable dishes.	Cook with other vegetables or sauté 3-5 minutes until translucent before adding to soup.	Incorporate into soups.	
Corn	Tear off husk and silk. Cut off the cob and serve in salads or salsas.	Heat oven to 350°. Place corn in the oven with the husk. Cook directly on oven rack for 25 minutes. Rub with butter, salt, and pepper.	Melt 1 tablespoon butter in a large sauté pan over medium heat. Sauté cut corn for – minutes. Season with 1 teaspoon chili powder, ½ teaspoon cumin, salt, and a squeeze of lime.	Place whole in pot of boiling water for 7 minutes.	Wrap clean corn cobs in moist paper towel and sprinkle with lemon juice. Cook for 5 minutes.
Eggplant	Rinse and cut off stem. Cut according to cooking method.	Slice in 1 inch rounds. Rub with oil, salt, and pepper, and bake for 20 minutes. Flip halfway through. Sprinkle with shredded Parmesan cheese.	Cut into cubes. Sauté in oil, covered, for ten minute stirring occasionally. Toss with salt, pepper, and Parmesan cheese.		
Greens Collards, Chard, Kale, Mustard, Turnip, Box Choy	Make sure to remove all dirt and sand. Cut into strips or tear into pieces.		Sauté onions and garlic. Add sliced greens and cook until tender, 3-5 minutes. Season with salt and pepper.	Steam in chicken or vegetable broth for 3-5 minutes. Season with onion and garlic powder. Add vinegar if desired.	Cook 8-10 minutes.

		Cooking Method			
Salad Greens arugula, lettuce, mesculine, spinach, kale.	Do not cut with a metal knife, leaves will turn brown. Tear or cut with silicon knife. Enjoy as a salad.				
Green Beans	Pop ends off and keep whole or cut into 1 inch sticks.	Toss with oil, garlic powder, salt, and pepper and roast for 15 minutes. Toss with 1 tablespoon red wine vinegar and 1 teaspoon honey.	Cook 4-6 minutes.	Steam 4-6 minutes.	Cook 8-10 minutes.
Herbs basil, chives, cilantro, dill, parsley	Lightly chop and add to salads, dressings, or at the end of cook time.				
Onions	Cut in half and peel off outer layers. Slice, chop, or mince and add to salads.	Toss with oil and roast with other vegetables.	Cook 5 minutes or on low for 10 minutes to caramelize.		
Peas	Shell		Cook in small amount of oil for 5-8 minutes.	Cook 5-8 minutes to maintain color and flavor.	Cook 3-5 minutes.
Peppers	Cut the top off. Pull out seeds. Cut into slices and enjoy raw or cut as desired and add to salads or salsas.	Toss with a small amount of oil. Keep whole after removing stem and seeds, rub with oil, and cook cut side down 20-25 minutes. Allow to cool and then remove skin for a roasted pepper.	Cook in oil 3-5 minutes and season with salt and pepper.	Only cut off the top. Place cut side down and cook, covered, 10-15 minutes.	Cook 5-7 minutes.
Radish	Slice and enjoy on salads or with dip.				

		Cooking Method			
Root vegetable beet, parsnip, potato, rutabaga	Peel and dice or slice.	Toss with oil, salt, and pepper and bake 20-25 minutes. To bake whole: pierce with a fork, wrap in foil, place on baking sheet and bake 25-60 minutes.	Cook in oil 10-15 minutes. Season with salt, pepper, and seasoning of choice.	Cover with water, bring to a boil, and cook 20-25 minutes. Toss with butter, garlic, and chives or onions. For mashed potatoes add a small amount of milk and whip.	Pierce whole root with a fork and wrap in a wet paper towel. Microwave 7-15 minutes.
Squash winter	Cut in half and remove seeds. Keep whole or peel and cube. *Clean seeds and roast for a snack.	Rub with oil. Place cut side down on a baking sheet lined with foil. Bake 45-55 minutes. Stuff with whole grains and vegetables or remove "meat" and add to pasta.	Cook cubes in oil 10-15 minutes. Add to pasta, vegetable stir fry, or add to salad once chilled.	Cover cubes with water and cook 20-25 minutes. Place cut side down with a small amount of water. Cook, covered, 20-40 minutes.	Place cut-side down on a dish with a small amount of water. Cook 15 minutes.
Zucchini, Summer Squash	Cut off ends. Slice into ¼ inch thick rounds or dice. Enjoy raw with dip.	Place on baking sheet lined with foil. Toss with oil, salt, and pepper OR favorite oil-based dressing. Cook 15-20 minutes.	Cook in oil with salt and pepper. Season with favorite seasoning blend or light oil-based dressing.	Incorporate into soups towards the end of cooking so as not to over cook.	Cook 8-10 minutes with seasoning and cherry tomatoes.

Arugula Topped Green Pizza

Prep time: 10 minutes
Cook time: 10 minutes
Makes 6 portions

INGREDIENTS:

- 1 pound prepared pizza dough, preferably whole wheat
- 2 cups chopped broccoli florets
- ¼ cup water
- 5 ounces Arugula, any tough stems removed, chopped about 6 cups
- Pinch of salt
- Freshly ground pepper to taste
- ½ cup prepared pesto
- 1 cup shredded part-skim mozzarella cheese

DIRECTIONS:

- Position oven rack in the lowest position; preheat to 450°F. Coat a large baking sheet with cooking spray.
- Roll out dough on a lightly floured surface to about the size of the baking sheet. Transfer to the baking sheet. Bake until puffed and lightly crisped on the bottom, 8 to 10 minutes.
- Meanwhile, cook broccoli and water in a large skillet over medium heat, covered, until the broccoli is crisp to tender, about 3 minutes. Stir in arugula and cook, stirring, until wilted, 1 to 2 minutes more. Season with salt and pepper.
- Spread pesto evenly over the crust, top with the broccoli mixture and sprinkle with cheese. Bake until crispy and golden and the cheese is melted, 8 to 10 minutes.

—Adapted from MUSC Urban Farm Crop Guide

Pea Greens Salad

Prep time: 20 minutes
Cook time: 0 minutes
Makes approximately 10, ¾ cup portions

INGREDIENTS:

- 2 ½ cups sugar snap peas
- 1 cup snow peas
- 1 cup green peas, fresh or frozen
- 4 tablespoons lemon juice
- 1 teaspoon honey
- 1 teaspoon salt
- ½ cup light olive oil
- 1 ½ cup radishes, sliced
- 2 heads lettuce
- 6 tablespoons mint

DIRECTIONS:

- In a pot of boiling salted water, cook sugar snap peas for 2 minutes.
- Add snow peas and green peas and cook 1 minute.
- Drain and "shock" in cold water and set aside in the refrigerator.
- In a medium bowl, whisk lemon juice, oil, salt, and honey.
- In a large bowl, toss all peas and radishes with dressing.
- Wash and separate lettuce leaves, pat dry and place on platter.
- Spoon salad onto lettuce leaves and garnish with chopped mint.

Pea Guacamole

Prep time: 10 minutes
Cook time: 4 minutes
Makes 4, ⅓ cup servings

INGREDIENTS:

- 1 lb frozen/fresh peas
- 1 small onion, grated
- 2 cloves garlic, crushed
- 1 red chili, de-seeded and chopped
- Juice of 1 lime
- 2 tbsp coriander, chopped
- Salt and freshly ground black pepper
- Tabasco

DIRECTIONS:

- Cook peas for 3-4 minutes in lightly salted boiling water or until tender. Drain and allow to cool slightly.
- Put peas in a food processor with all remaining ingredients apart from a sprinkling of the chopped chili. Pulse until mixture is crushed but not completely smooth.
- Season with salt, freshly ground black pepper and a few drops of Tabasco as required.
- Transfer to a serving bowl and garnish with remaining chopped chili.

—Adapted from MUSC Urban Farm Crop Guide

Mashed Cauliflower

Prep time: 15 minutes
Cook time: 10 minutes

INGREDIENTS:

- 1 head of cauliflower (about 8 cups)
- 1 tablespoon cream cheese
- ½ teaspoon salt
- Black pepper to taste
- 2 garlic cloves, minced
- 1 tablespoon butter

DIRECTIONS:

- Place a large pot of water over high heat. Meanwhile, wash and cut the head of cauliflower into bite-size pieces.
- Once the water comes to a boil, put the cauliflower into the pot. Cook for about 10 minutes, or until very tender.
- Drain the cauliflower and pat dry with paper towels.
- For a chunkier dish: add the cauliflower to a large bowl with the cream cheese, salt, pepper, garlic, and butter. Mash well with a potato masher or fork, until no large pieces are left.
- For a smoother dish: Add the cauliflower to a blender or food processor with the cream cheese, salt, pepper, garlic, and butter. Blend until smooth.

—Adapted from the Greater Pittsburgh Food Bank

Eggplant Casserole

Prep time: 15 minutes
Cook time: 45 minutes

INGREDIENTS:

- 1 large onion, chopped
- 1 large green pepper, chopped
- 1 cup chopped celery
- 4 tablespoons butter
- 1 medium-large eggplant
- 1 cup fresh bread crumbs
- Salt and Pepper to taste
- 2 eggs, beaten
- 1 cup skim milk
- 2 teaspoons Worcestershire sauce

DIRECTIONS:

- Sauté the onion, pepper, and celery in butter.
- Peel and cut the eggplant into 1 inch cubes. Boil until tender.
- Add eggplant to mixture. Add bread crumbs, salt, pepper, eggs, milk, and Worcestershire sauce.
- Place all ingredients in a buttered dish. Bake at 350°F for 30-45 minutes.

—*Adapted from The Green Tomato Patch Cookbook*

Parmesan Roasted Broccoli

Prep time: 5 minutes
Cook time: 20-25 minutes
Makes approximately 4, 1 cup portions

INGREDIENTS:

- 5 lbs trimmed broccoli (9 cups of florets)
- 1 tablespoon minced garlic
- 5 tablespoons olive oil
- 2 tablespoons freshly squeeze lemon juice
- 4 tablespoons pine nuts, toasted
- ¼ cup freshly grated Parmesan cheese
- ¼ cup freshly grated Romano cheese
- 6 basil leaves, sliced julienned
- Season with salt and pepper

DIRECTIONS:

- Preheat the oven to 425°F.
- Place the broccoli florets in a single layer on a sheet pan.
- Toss the garlic on the broccoli and drizzle the olive oil, season with salt and pepper.
- Roast for 20 to 25 minutes, until crispy and tender, some of the florets may be browned.
- Immediately toss roasted broccoli with 2 tablespoons olive oil, lemon juice, Parmesan cheese, Romano cheese, and basil.

—Adapted from MUSC Urban Farm Crop Guide

Cabbage-Apple Slaw

Prep time: 15 minutes
Cook time: 8-10 minutes
Makes approximately 4, ¾ cup portions

INGREDIENTS:

- 3 cups cabbage green or purple
- 2 medium apples
- ¼ cup water
- 1 tablespoon apple cider vinegar
- A pinch of salt

DIRECTIONS:

- Shred cabbage.
- Cut apples into thin slices.
- In a medium skillet, cook cabbage, apple, and water covered. Cook until tender.
- Stir in vinegar and salt.
- Serve warm as a side or topping of a meat dish.

—Adapted from MUSC Urban Farm Crop Guide

White Bean and Kale Soup

Prep time: 15 minutes
Cook time: 40 minutes
Makes approximately 10, ¾ cup portions

INGREDIENTS:

- 1 ½ teaspoon olive oil
- 10 cloves garlic, minced
- 1 onion, chopped
- 5 cups kale, chopped
- 5 cup low sodium chicken broth
- 2 cans (5 cups) white beans, drained and rinsed
- 2 cups plum tomatoes, diced
- 2 teaspoons Italian seasoning
- 1 ½ cup parsley

DIRECTIONS:

- In a large pot, heat olive oil. Add garlic and onion; sauté until soft.
- Add kale and sauté, stirring until wilted.
- Add 3 cups of broth, 2 cups of beans, and all of the tomatoes, herbs, salt, and pepper. Bring to a boil and simmer 5 minutes.
- In a blender or food processor, mix the remaining beans and broth until smooth.
- Stir pureed beans into soup to thicken. Bring back to a boil and simmer 15 minutes.
- Ladle into bowls and serve with chopped parsley.

Spaghetti Squash with Basil and Parmesan

Prep time: 5 minutes
Cook time: 45-60 minutes

INGREDIENTS:

- 1 ½ pound spaghetti squash
- ¼ cup freshly grated Parmesan cheese
- 1 tablespoon fresh basil, chopped
- ¼ teaspoon salt
- ¼ teaspoon pepper

DIRECTIONS:

- Preheat oven to 350°F.
- Cut squash in half, lengthwise, and remove seeds. Place cut side down on an oiled sheet pan. Bake for 45-60 minutes or until fork tender. Once cooled enough to touch, shred squash flesh with a fork.
- Add Parmesan cheese, basil, salt, and pepper. Toss evenly to coat.

—Adapted from MUSC Urban Farm Crop Guide

Roasted Vegetable Soup

Prep Time: 20 minutes
Cook Time: 50-60 minutes
Makes approximately 10, ⅓ cup servings

INGREDIENTS:

- 2 lbs carrots, peeled
- 1 lb onions, peeled
- 2 lbs parsnips, peeled
- 2 sweet potato, peeled
- 3 lbs butternut squash, peeled and seeded
- ⅓ cup olive oil
- 1 tablespoon salt
- 1 teaspoon black pepper
- ¼ cup parsley
- ½ gallon chicken or vegetable stock

DIRECTIONS:

- Preheat oven to 425°F.
- Cut all vegetables into 1 inch cubes.
- Coat all cut vegetables with olive oil and toss in a bowl with salt and pepper.
- Place in oven and roast until tender, about 25-35 minutes, turning vegetables once.
- Place the roasted vegetables into a stock pot with chicken or vegetable stock.
- Simmer on stove 20 minutes and season to taste with salt and pepper.

—Adapted from MUSC Urban Farm Crop Guide

Parsnip and Carrot Fries

Prep time: 20 minutes
Cook time: 40 minutes

INGREDIENTS:

- 2 pounds parsnips, peeled
- 1 pound carrots, un-peeled
- 3 tablespoons olive oil
- 1 tablespoon salt
- 1 ½ teaspoon ground black pepper

DIRECTIONS:

- Preheat the oven to 425°F.
- Cut the parsnips and carrots into fries. Don't make them too small. Place on a sheet pan covered in foil. Toss with olive oil and sprinkle with salt and pepper. Roast for 20-40 minutes depending on the size of the fries. Toss halfway through.

—Adapted from Ina Garten

Asparagus with Lemon Garlic Vinaigrette

Prep time: 5 minutes
Cook time: 5 minutes
Makes approximately 4, ½ cup servings

INGREDIENTS:

- 1 pound fresh asparagus, washed and trimmed
- 1 fresh garlic clove, chopped
- ¼ teaspoon salt
- 1 ¼ tablespoon fresh lemon juice
- 1 teaspoon Dijon mustard
- 3 tablespoons olive oil
- ¼ teaspoon sugar
- ¼ teaspoon pepper
- Fresh parsley, chopped, for garnish

DIRECTIONS:

- In a food processor, combine garlic and salt for dressing. Process until well blended.
- Add lemon juice and mustard. Process to a smooth paste.
- Add oil in a slow, steady stream until incorporated.
- Add sugar and pepper. Mix well.
- Place asparagus in steamer and steam for 3 minutes. Immediately plunge asparagus into ice water. Drain well. Pat dry with paper towels if necessary. Dressing will adhere better if the asparagus is not wet.
- Arrange asparagus on chilled serving platter not stainless with stems facing the same direction. Drizzle with dressing. Gently toss asparagus with tongs to evenly coat.
- Sprinkle with parsley.

Notes: Try different dressings such as balsamic vinaigrette or sprinkle with shredded cheese.

—Adapted from MUSC Urban Farm Crop Guide

Balsamic Dressed Roasted Beets

Prep time: 10 minutes
Cook time: approximately 1 hour
Makes approximately 8, ½ cup portions

INGREDIENTS:

- 6 medium beets about 2 ½ pounds
- ½ cup fresh orange juice
- ¼ cup balsamic vinegar
- 1 tablespoon sugar
- ½ teaspoon salt
- ⅛ teaspoon freshly ground black pepper

DIRECTIONS:

- Preheat oven to 400°.
- Leave root and 1 inch of stem on beets; scrub with a brush. Wrap beets in foil. Bake at 400° for 1 hour or until tender. Cool beets to room temperature.
- Peel and cut each beet into 8 wedges.
- Combine juice, vinegar, and sugar in a small saucepan; bring to a boil. Cook until reduced to ⅓ cup about 10 minutes.
- Combine beets, vinegar mixture, salt, and pepper; toss well.

—Adapted from MUSC Urban Farm Crop Guide

Carrot Salad

Prep time: 5 minutes
Cook time: 0 minutes
Makes approximately 4, ¼ cup servings

INGREDIENTS:

- 2 cups fresh carrots, sliced ⅛ inch thick
- 1 ½ tablespoon honey
- 1 ½ tablespoon fresh parsley, chopped
- 3 tablespoons light raspberry vinaigrette
- 3 tablespoons apple cider

DIRECTIONS:

- In a medium bowl, combine honey, parsley, raspberry vinaigrette, and apple cider.
- Mix sliced carrots with dressing.
- Keep in refrigerator for at least 2 hours before serving.

Notes: Keeps in the refrigerator for 2-5 days.

—Adapted from MUSC Urban Farm Crop Guide

Kale Slaw with Red Cabbage and Carrots

Prep time: 10 minutes
Cook time: 0 minutes
Makes approximately 4 servings

INGREDIENTS:

- 1 tablespoon olive oil
- 1 tablespoon Dijon mustard
- 1 teaspoon apple-cider vinegar
- Salt and pepper
- 1 ½ cups kale
- 1 ½ cups red cabbage
- 1 carrot, grated
- ¼ cup fresh parsley leaves
- 2 tablespoons red onion, diced
- 2 tablespoons sunflower seeds
- 2 tablespoons pumpkin seeds

DIRECTIONS:

- In a small bowl, whisk olive oil, mustard, and apple-cider vinegar. Season with salt and pepper.
- Shred kale and cabbage.
- In another bowl, combine kale, cabbage, carrot, parsley, and red onion with sunflower and pumpkin seeds.
- Season with salt and pepper, drizzle with dressing, and toss to coat.

—Adapted from MarthaStewart.com

Grilled Pepper Salad

Prep time: 10 minutes
Cook time: 10 minutes
Makes approximately 4, 1 cup portions

INGREDIENTS:

- 4 bell peppers, mixed colors , halved, seeded and stemmed
- 1 small purple onion, sliced in rounds
- ¼ cup rinsed and chopped oil-packed sun-dried tomatoes
- 1 tablespoon extra virgin olive oil
- 1 tablespoon balsamic vinegar
- ⅛ teaspoon salt
- Optional: ¼ cup halved and pitted black olives

DIRECTIONS:

- Place peppers and onions on foil. Grill on medium-high heat. Turn once, until soft and charred in spots, about 5 minutes per side.
- When cool enough to handle, chop the peppers and onions; toss with olives, sun-dried tomatoes, oil, vinegar and salt in a large bowl.

Notes: Bake peppers and onions at 400°F for 10 minutes, flipping halfway through cook time.

—Adapted from MUSC Urban Farm Crop Guide

Sautéed Spinach with Red Onion and Garlic

Prep time: 2 minutes
Cook time: 6 minutes

INGREDIENTS:

- 1 lb fresh spinach
- 2 ½ teaspoons fresh garlic, minced
- ¼ cup fresh red onions, diced
- 1 ¼ teaspoon olive oil
- ⅓ teaspoon salt
- ⅓ teaspoon ground black pepper

DIRECTIONS:

- In a medium sauté pan, heat olive oil over medium heat.
- Add garlic and red onions and sauté for 2 minutes or until onions are translucent. Add spinach, stir and sauté until spinach is tender.
- Stir in salt and pepper. Serve immediately.

— *Adapted from MUSC Urban Farm Crop Guide*

Soulful Greens

Prep time: 10 minutes
Cook time: 30 minutes

INGREDIENTS:

- ½ cup reduced sodium chicken broth
- ¾ cup water
- 2 pounds collard greens, washed, stems removed, and chopped
- 1 ½ cups red onions, sliced
- 1 clove garlic, minced
- ¼ cup orange juice
- ½ teaspoon dried red pepper flakes

DIRECTIONS:

- Pour chicken broth and water in a large pot. Bring to a boil. Add collards. Cook uncovered at medium-high heat for 15 minutes. Drain the collards.
- Sauté garlic and onions for 5 minutes in a large nonstick skillet. Add orange juice and wilted greens.
- Stir until well coated. Simmer uncovered at medium-low for 10-15 minutes.
- Sprinkle with red pepper flakes and serve.

— *Adapted from Eat Smart Move More North Carolina*

Black Eyed Pea Salad

Prep time: 10-15 minutes
Cooking time: 40-50 minutes
Makes approximately 10, ¾ cup portions

INGREDIENTS:

- 5 cups black eyed peas, fresh cooked, canned or frozen
- 3 cups cherry tomatoes, quartered
- 5 peaches, seed removed, diced
- 1 red onion, minced
- 1 lime, juiced
- 1 ½ tablespoons olive oil
- ¼ cup cilantro leaves, chopped
- 2 tablespoons parsley leaves, chopped
- ½ tablespoon cumin, ground
- Salt and pepper

DIRECTIONS:

- Mix all ingredients together in a stainless steel bowl. Let chill in the refrigerator for about an hour to allow flavors to blend.

Breakfast Quinoa with Blueberries

Prep time: 2 minutes
Cook time: 23 minutes
Makes approximately 2, ½ cup portions

INGREDIENTS:

- 2 cups low-fat milk
- 1 cup quinoa, rinsed
- 3 tablespoons light brown sugar
- ⅛ teaspoon ground cinnamon
- 1 cup ½ pint fresh blueberries

DIRECTIONS:

- Bring milk to a boil in a small saucepan. Add quinoa and return to a boil. Reduce heat to low, simmer covered until three quarters of the milk is absorbed, about 15 minutes.
- Stir in sugar and cinnamon. Cook covered until most of the milk is absorbed, about 8 minutes.
- Stir in blueberries and cook 30 seconds. Serve with extra milk, sugar, cinnamon, and blueberries.

—Adapted from MUSC Urban Farm Crop Guide

Mini Caprese Salad

Prep time: 15 minutes
Cook time: 0 minutes
Makes approximately 10, ½ cup servings

INGREDIENTS:

- 4 cups cherry tomatoes, halved
- 2 cups mozzarella balls bocconcini
- 1 ½ cups basil, fresh
- ½ cup extra virgin olive oil
- 2 cloves garlic
- 1 lemon, juiced
- Salt and pepper

DIRECTIONS:

- In a food processor, combine lemon juice, garlic, and basil.
- While the food processor is running, slowly pour in olive oil to form a smooth dressing.
- In a large bowl, combine tomatoes, cheese, and dressing.
- Salt and pepper to taste and serve.

Notes: If you do not have a food processor, mince garlic and basil and mix with lemon juice. Slowly whisk oil into lemon juice, beating rapidly to incorporate all of the ingredients.

Strawberry and Spinach Salad

Prep time: 10 minutes
Cook time: 0 minutes
Makes approximately 10, ¾ cup servings

INGREDIENTS:

- 1 cup strawberries
- 10 ounces baby spinach, washed and dried
- ⅓ cup feta cheese, crumbled
- ⅓ cup almonds, slivered

Dressing:

- ¼ cup strawberries, fresh, sliced, and hulled
- 3 tablespoons white wine vinegar
- 3 tablespoons basil leaves, fresh, chopped
- 1 ½ tablespoon olive oil
- 1 teaspoon honey
- Salt and pepper

DIRECTIONS:

- Place strawberries, white wine vinegar, basil, honey, salt, and pepper in a food processor or blender. If you don't have a food processor, dice or mince strawberries and basil and whisk together with ingredients.
- Toss spinach, cheese, almonds, and strawberries in a large bowl.
- Lightly drizzle with dressing and serve.

Sweet Sour Watermelon and Cucumber Salad

Prep time: 20 minutes
Cook time: 0 minutes
Makes approximately 10, ½ cup portions

INGREDIENTS:

- 4 cups watermelon, cubed
- 2 cucumbers, peeled and sliced
- 3 scallions, chopped
- ½ cup cider vinegar
- 1 cup water
- ⅓ cup honey
- 1 tablespoon poppy seeds
- Salt
- Pepper

DIRECTIONS:

- Mix the watermelon and cucumber in a bowl.
- In a small bowl, whisk vinegar, water, and honey.
- Stir in scallions and poppy seeds.
- Season with salt and pepper.
- Toss the salad with just enough dressing to coat. Cover and refrigerate.

Baked Zucchini

Prep time: 5 minutes
Cook time: 15 minutes

INGREDIENTS:

- 4 medium zucchini, sliced into ½ inch rounds
- 2 ounces part skim mozzarella cheese, shredded
- 1 teaspoon garlic powder
- 1 tablespoon parsley, chopped

DIRECTIONS:

- Preheat oven to 350°F.
- Cover the bottom of a baking dish with half the zucchini. Sprinkle with half of the cheese, garlic powder, and parsley. Repeat with a second layer using the remaining ingredients.
- Bake until cheese is melted and lightly golden, about 10 to 15 minutes.

—Adapted from Eat Smart Move More NC Celebrating the Harvest

SOUTH CAROLINA SEASONAL RAINBOW

Winter Rainbow

Red	Yellow/Orange	Tan/White	Green	Blue/Purple
	carrots, parsnip, winter squash	mushrooms, rutabaga, turnip	arugula, broccoli, cabbage, chard, cilantro, collards, kale, lettuce, mustard greens, scallions, spinach	beets

Spring Rainbow

Red	Yellow/Orange	Tan/White	Green	Blue/Purple
red cabbage, red onion, tomatoes	sweet potatoes, carrots	radish, potatoes, onions	asparagus, bok choy, cabbage, chard, cilantro, collards, dill, kale, lettuce, mesculine mix, mustard greens, parsley, peas, spinach	beets

Summer Rainbow

Red	Yellow/Orange	Tan/White	Green	Blue/Purple
raspberries, red peppers, strawberries, tomatoes, watermelon	cantaloupe, corn, yellow/orange peaches, pears, peppers, squash blossoms, summer squash	garlic, ginger	basil, celery, chives, cucumbers, edamame, green onions, leeks, peppers, zucchini	blackberries, eggplant, figs, muscadine grapes

Fall Rainbow

Red	Yellow/Orange	Tan/White	Green	Blue/Purple
pomegranates	citrus, persimmon, pumpkin	beans, cauliflower, onions, turnip	arugula, bok choy, chives, dill, kale, kiwi, lettuce, parsley, peas	eggplant

VOCABULARY

1In Merriam-Webster Dictionary online. Retrieved from http://www.merriam-webster.com.

2Zais, Mike, 2014 "Support Documents for South Carolina Academic Standards and Performance Indicators for Science", Grades K-8, Biology, and Earth Science. Available from http://ed.sc.gov/agency/se/Instructional-Practices-and-Evaluations/Science.cfm.

WORD	TYPE OF SPEECH	DEFINITION
Abiotic	adj.	Not biotic; Not involving or produced by organisms.1 The non-living parts of the ecosystem.2
Adaptations	n.	A change in a plant or animal that makes it better able to live in a particular place or situation.1 A trait or behavior that helps an organism survive and reproduce. 2
Adult	adj.	Fully grown and developed.1
Anemometer	n.	An instrument for measuring and indicating the force or speed and sometimes direction of the wind.1
Animal	n.	Any living thing that is not a plant.1 Multicellular organisms whose cells have nuclei. Almost all move to get food, and are heterotrophs. 2
Annual	n.	An annual is a plant that completes its life cycle in one growing season. It will grow, flower, set seed, die, and decompose.1
Anterior	adj.	Near or toward the front of something, such as the body. 1
Anthocyanins	n.	Any of a class of water-soluble pigments that give flowers the colors ranging from red to blue. 1
Anthoxanthins	n.	Any of a class of water-soluble pigments that give flowers the colors ranging from white or colorless to a creamy to yellow. 1
Apron	n.	A piece of clothing that is worn on the front of the body over clothes to keep them from getting dirty.1
Attract	v.	Cause to come to a place or participate in a venture by offering something of interest, favorable conditions, or opportunities. 1
Axis	n.	The imaginary straight line that something, such as the Earth, turns around.1 Earth rotates around an imaginary straight line that runs through the planet s center. 2

Bacteria	n.	Group of tiny, single-celled organisms that inhabit virtually all environments, including soil, water, organic matter, and the bodies of multi cellular animals. 1
Barometer	n.	An instrument that is used to measure air pressure and predict changes in the weather.1
Biennial	n.	A biennial is a plant that needs two growing seasons to complete its life cycle. It produces leaves one season. Then it goes dormant or rests over the winter. In the spring, it will begin to grow again and grow flowers, set seed, die, and decompose. The seed that is left behind on the ground germinates and the cycle begins again. 1
Biotic	adj.	Of, relating to, or caused by living organisms.1
Broad	adj.	Large from one side to the other side.1
Bumpy	adj.	Of a surface: having or covered with bumps.1
Burrow	n.	A hole or tunnel in the ground that an animal, such as a rabbit or fox, makes to live in or for safety.1
Byproduct	n.	Something that is produced during the production or destruction of something else.1
Calcium	n.	A substance that is found in most plants and animals and that is especially important in people for strong healthy bones.1
Carbon Dioxide	n.	A gas that is produced when people and animals breathe out or when certain fuels are burned and that is used by plants for energy.1
Carteniods	n.	Any of a class of fat-soluble pigments that give flowers the colors ranging from yellow to orange to red. 1
Cells	n.	The smallest structural and functional unit of an organism consisting of cytoplasm and a nucleus enclosed in a membrane. 1 The basic units of structure and function within all living things. 2
Chisel	n.	A metal tool with a flat, sharp end that is used to cut and shape a solid material such as wood, stone, or metal. 1
Chlorophyll	n.	The green substance in plants that makes it possible for them to make food from carbon dioxide and water. 1 A green pigment that absorbs light energy from the sun. 2
Chloroplast	n.	A cellular part that contains chlorophyll and is the location of photosynthesis. 1 Found in the cells of the leaf containing chlorophyll, and are the sites where photosynthesis takes place in a plant cell. 2

Circular	adj.	Shaped life a circle or part of a circle. 1
Cinquain	n.	A 5-line stanza. 1
Clay	n.	A stiff, sticky fine-grained earth, typically yellow, red, or bluish-gray in color and often forming an impermeable layer in the soil. It can be molded when wet, and is dried and baked to make bricks, pottery, and ceramics. 1 A type of soil that has very small grains, much smaller than sand or silt, and holds water easily. This makes it sticky when wet, but when it dries, it forms hard clumps. 2
Clitellum	n.	The wide band found on the body of a worm. 1
Cloud	n.	A white or grey mass of particles of condensed vapor water or ice in the sky. 1
Cloudy	adj.	Having many clouds in the sky. 1
Coexist	v.	To exist together or at the same time. 1
Collect	v.	To gather an accumulation of objects. 1
Composition	n.	The way in which something is put together or arranged. 1
Compost	n.	A decayed mixture of plants, such as leaves and grass, that is used to improve the soil in a garden. 1
Condensation	n.	The art or process of condensing, a chemical reaction involving union between molecules often with elimination of a simple molecule, like water, to form a new more complex compound of often greater molecular weight. 1 The process of water vapor changing back to droplets of water, as a result of the cooling air temperature. 2
Conservation	n.	The careful use of natural resources, such as trees, oil, to prevent them from being lost or wasted. 1
Conserve	v.	To keep safe from being damaged or destroyed. 1
Cotyledon	n.	The first leaf or one of the first pair or whorl of leaves developed by the embryo of a seed plant or of some lower plants.1
Creature	n.	An animal of any type.1
Cup	n.	A small round container that often has a handle and that is used for drinking or measuring liquids.1
Dairy	n.	Food produced from the milk of mammals.1
Debris	n.	The pieces that are left after something has been destroyed.1

Decay	v.	To be slowly destroyed by natural processes that destroy a dead plant.1
Decompose	v.	To cause something, such as dead plants and the bodies of dead animals, to be slowly destroyed and broken down by natural processes, chemicals, etc.1
Decomposer	n.	Any various organisms, such as bacteria and fungi, that return constituents of organic substances to ecological cycles by feeding on and breaking down dead protoplasm.1 Consumers, including microorganisms, termites, worms, and fungi, that get the energy they need by breaking down dead or decaying matter. 2
Decrease	v.	To become smaller in size, amount, number, etc.1
Defense	n.	A mechanism or method that is used to protect an organism.1
Depth	n.	A part that is far from the outside or surface.1
Detect	v.	To discover or notice the presence of.1
Dew Points	n.	The temperature at which the moisture in the air forms visible drops of water.1
Diameter	n.	A straight line from one side of something, such as a circle, to the other side that passes through the center point.1
Digest	n.	Information or a piece of writing that has been made shorter.1
Direct Sunlight	n.	Light directly from the sun.1
Dormant	adj.	Not active but able to become active.1
Earth	n.	The planet on which we live, or the material in which plants grow.1 The third planet from the Sun in the solar system. It has a rocky surface and contains water. 2
Ecosystem	n.	Everything that exists in a particular environment.1 A community all the organisms in a given area and the abiotic factors such as water, soil, or climate that affect them. 2
Edible	adj.	Suitable or safe to eat.1
Embryo	n.	A human or animal in the early stages of development before it is born, hatched, etc.1
Endangered	adj.	Being or relating to an endangered species.1
Energy	n.	A fundamental entity of nature that is transferred between parts of a system in the production of physical change whiten the system and usually regarded as the capacity for doing work.1 The ability to make something move, happen, or change. 2

Engineer	n.	A person who has scientific training and who designs and builds complicated products, machines, systems, or structures.1
Entrance	n.	The way in which one enters something.1
Environment	n.	The conditions and influences that affect the growth, health, progress, etc., of someone or something.1
Equator	n.	The great circle of the celestial sphere whose plane is perpendicular to the axis of the earth.1
Erosion	n.	The gradual destruction of something by natural forces, such as water, wind, or ice.1 When earth materials, like rock, sand, and soil, are being carried away from their original location, often caused by water or wind. 2
Evaporation	v.	To change from a liquid into a gas.1 A change in state from a liquid to a gas, and occurs at the surface of the liquid as heat is added from the surroundings. This process results from the sun s energy. 2
Examine	v.	To look at something closely and carefully in order to learn more about it, to find problems, etc.1
Exist	v.	To continue to be or to live.1
Experiment	n.	A scientific test in which you perform a series of actions and carefully observe their effects in order to learn about something.1
Explore	v.	To look at something in a careful way to learn more about it.1
Extinct	adj.	No longer existing.1
Feast	n.	A special, formal meal with large amounts of food and drink.1
Fertile	adj.	Able to support the growth of many plants.1
Fertilize	v.	To make a plant or flower able to produce seeds.1
Fewer	adj.	A smaller number of persons or things.1
Flat	adj.	Having a smooth, level, or even surface.1
Flower	n.	The part of a plant that is often brightly colored, that usually lasts a short time, and from which the seed or fruit develops.1 The part of the flowering plant that helps the plant make new plants like itself. It has special characteristics, such as color or scent, which attracts insects and makes the seeds. 2
Foggy	adj.	Having or filled with fog.1

Food Chain	n.	A series of types of living things in which each one uses the next lower member of the series as a source of food.1 A series of plants and animals in which each organism is a source of food energy for the next in the series. 2
Food Web	n.	The totality of interacting food chains in an ecological community.1 A diagram describing the complex patterns of energy flow in an ecosystem by modeling who consumes what. It describes the organisms in a particular ecosystem found in interconnecting food chains. 2
Friction	n.	The act of rubbing one thing against another.1 The rubbing of two objects together. 2
Fruit	n.	The part of a plant that has the seeds in it, such as the pod of a pea, a nut, a grain, or a berry.1 The part of the flowering plant that grows from the flower and contains the seeds. They are formed around the seed to protect it. 2
Garbage	n.	Things that are no longer useful or wanted and that have been thrown out.1 Geotropism
Germination	v.	To cause a seed to begin to grow.1 The process in which a plant begins to sprout or grow from the seed. 2
Glove	n.	A covering for the hand that has separate parts for each finger.1
Habitat	n.	The place or type of place where a plant or animal naturally or normally lives or grows.1
Half	n.	One of two equal parts into which something can be divided.1
Hand Shear/Pruner	n.	A small tool used to reduce something by removing parts that are not necessary or wanted.1
Hand Span	n.	The distance from the tip of the thumb to the tip of the little finger on your outstretched hand.1
Hand Trowel	n.	A small tool with a curved blade that is used by gardeners for digging holes.1
Harmful	adj.	Causing or capable of causing damage or harm.1
Harvest	n.	The season when crops are gathered from the fields or the activity of gathering crops.1
Helpful	adj.	Making it easier to do a job, deal with a problem, etc.1
Hoe	n.	A garden tool that has a flat blade on a long handle.1
Hooked	adj.	Shaped like a hook.1

Humus	n.	A brown or black material in soil that is formed when plants and animals decay.1 Soil that is made up of decayed parts of once-living organisms, and is dark, soft, and very crumbly. 2
Immune system	n.	The system that protects your body from diseases and infections.1
Increase	v.	To become larger or greater in size, amount, number, etc.1
Insect	n.	A small animal with well-defined head, thorax, and abdomen, only three pairs of legs, and typically one or two pairs of wings.1
Intestines	n.	The tubular part of the alimentary canal in the body that helps digest food after it leaves the stomach.1
Invasive	adj.	Involving entry into a body by cutting or by inserting an instrument.1
Investigate	v.	To try to find the facts about something in order to learn how it happened.1
Large	adj.	Great in size, not limited in importance, range, etc.1
Leaf	n.	One of the flat and typically green parts of a plant that grows from a stem or twig.1 The flat, usually green, part of the plant that grows from the stem and makes food for the plant. 2
Least	adj.	Smallest in amount or degree.1
Leaves	pl. n.	One of the flat and typically green parts of a plant that grow from a stem or twig.1 The part of the plant that produce food in the presence of light. 2
Light	n.	The brightness produced by the sun, by fire, a lamp, etc.1
Locomotion	n.	The act or power of moving from place to place.1
Lunar	adj.	Of or relating to the moon.1
Lycopene	n.	A specific red carteniod pigment present in tomatoes and many berries and fruits.1
Magnetic Field	n.	The portion of space near a magnetic body or a body carrying an electric current within which magnetic forces due to the body or current can be detected.1
Magnetize	v.	To cause something to become magnetic.1
Manmade	adj,	Made by people rather than by nature.1
Material	adj.	Relating to or made of matter.1

Mature Plant	n.	Having a fully grown or developed plant.1 Having the same structures as seedlings, but in addition, they are able to reproduce using flowers or cones. 2
Measurement	n.	The size, length, or amount known by measuring something.1
Micronutrients	n.	An organic compound essential in small amounts to the growth and health of an animal.1
Moisture	n.	A small amount of a liquid, such as water, that makes something wet or moist.1
Molecules	n.	The smallest possible amount of a particular substance that has all the characteristics of that substance.1
Moon	n.	The large round object that circles the earth and that shines at night by reflecting light from the sun.1 The natural satellite that orbits the earth, it has a rocky, dusty surface with many craters and no water or atmosphere. 2
More	adj.	Greater in amount, number or size.1
Most	adj.	Almost all, or the majority of.1
Motion	n.	An act, process, or instance of changing place.1
Natural	adj.	Existing in nature and not made or caused by people.1
Nectar	n.	A sweet liquid produced by plants and used by bees in making honey, or a thick juice made from a particular fruit.1
Nibbled Leaf	v.	To eat slowly or with small bites on a leaf.1
Nitrogen	n.	A chemical that has no color or smell and that makes up a large part of the atmosphere.1
Nocturnal	adj.	Active mainly during the night.1
Nonstandard Units of Measure	n.	Objects of the same size or length used to measure, but are not universally accepted, and are unknown outside local context.1
Notched	n.	A small cut that is shaped like a V and that is made on an edge or surface.1
Nutrients	n.	A substance that plants, animals, and people need to live and grow.1
Obtain	v.	To gain or get something.1
Oil	n.	Any of numerous unctuous combustible substances that are liquid or can be liquefied easily on warming, are soluble in ether but not in water, and leave a greasy stain on paper or cloth.1

Organic	adj.	Describing food grown or made without the use of artificial chemicals.1
Organic Material	n.	Food grown or made without the use of artificial chemicals.1
Organism	n.	An individual living thing.1
Oxygen	n.	A chemical that is found in the air, that has no color, taste, or smell, and that is necessary for life.1
Parent Material	n.	The underlying geological material in which soil horizons form.1
Particle Size	adj.	The diameter or volume of the grains in a sediment or sedimentary rock.1
Partly Cloudy	adj.	Somewhat but not completely clouded.1
Perennial	n.	A plant that lives for 3 or more years, and can grow, flower, set seed for many years, die, and decompose.1
Perishable	adj.	Likely to spoil or decay quickly, not likely to stay fresh for a long time if not eaten or used.1
Permeability	n.	The state or quality of a material or membrane that causes it to allow liquids or gases to pass through it.1
pH	n.	A measure of acidity and alkalinity of a solution that is a number on a scale 0 to 14. Lower numbers indicate more acidic, higher numbers indicate more alkaline.1 A measure of the acidity of a solution. 2
Phase/s	n.	The shape of the part of the moon that is visible at different times during a month.1
Phosphorous	n.	A poisonous chemical element that glows in the dark and burns when it is touched by air.1
Photosynthesis	n.	The process by which a green plant turns water and carbon dioxide into food when the plant is exposed to light.1 The process by which plants make their own food using energy from sunlight. 2
Phytonutrients	n.	A bioactive plant-derived compound associated with positive health effects.1
Plant Waste	n.	Products that are created through photosynthesis and not used by a plant, such as oxygen, that exists the plant through stomata, root cell walls, and other routes.1
Pollination	n.	The transfer of pollen from another to the stigma in angiosperms or from the microsporangium to the micropyle in the gymnosperm.1

Pollinator	n.	One that pollinates.1
Population	n.	A group of people or animals of a particular kind that live in a place.1 All members of one kind of organism that live in a particular area. 2
Potassium	n.	A soft, silver-white metal that is used especially in farming and industry.1
Precipitation	n.	The quality or state of being precipitate.1 A type of weather condition that is measured by a rain gauge and can be described as rain, snow, sleet, or hail. 2
Prefer	v.	To like better than someone or something else.1
Pressure	n.	The weight or force that is produced when something presses or pushes against something else.1
Prickly	adj.	Having many sharp points.1
Problem	n.	Something that is a source of trouble, worry, etc.1
Process	n.	A series of actions that produce something or that lead to a particular result.1
Produce	n.	Fresh fruits and vegetables.1
Protein	n.	A substance found in food that is an important part of the human diet.1 Molecules composed of chains of amino acids, and are involved in almost every function in the human body. 2
Quarter	n.	One of four equal parts of something.1
Rain Gauge	n.	An instrument for measuring the quantity of precipitation or rain.1
Rainy	adj.	Having a lot of rain.1
Rake	n.	An tool equipped with projecting prongs to gather material, such as leaves, or for loosening or smoothing the surface of the ground.1
Reaction	n.	The way someone acts or feels in response to something that happens, is said, etc.1
Recycle	v.	To make something new from something that has been used before.1
Reduce	v.	To make something smaller in size, amount, number, etc.1
Regenerate	adj.	Restored to a better, higher, or more worthy state.1

Renewable	adj.	Able to be extended for another time period.1
Replenish	v.	To fill or build up again.1
Reproduce	v.	To produce something that is the same as or very similar to it's original.1
Reproduction	n.	The process that produces babies, young animals, or new plants.1
Resistance	n.	The effort made to stop or to fight against someone or something.1
Resource	n.	A supply of that someone has and can use when it is needed.1
Respiration	n.	The physical and chemical processes by which an organism supplies its cells and tissues with elements needed for metabolism.1 The process by which plants break down the sugar in the cells to obtain energy. 2
Respond	n.	An engaged pillar supporting an arch or closing a colonnade or arcade.1
Reuse	v.	To use again.1
Revolution	n.	The action of moving around something in a path that is similar to a circle.1 The movement of Earth as it makes an orbit around the Sun in one year. 2
Root/s	n.	The part of a plant that grows underground, gets water from the ground, and holds the plant in place.1 The part of the plant that grows under the ground, holds the plant in place, and takes in water and nutrients from the ground. 2
Rotation	n.	The act or process of moving or turning around a central point.1
Round	adj.	Shaped like a circle or ball, having curves rather than angles.1
Runoff	n.	Water from rain or snow that flows over the surface of the ground into streams.1
Salad	n.	A mixture of raw green vegetables, such as different types of lettuce, usually combined with other raw vegetables, raw or cooked food, and combined with dressing.1
Salad Spinner	n.	Usually a large fork and spoon connected and used to mix salad ingredients.1
Sample	n.	A small amount of something that gives you information about the thing it was taken from.1

Sand	n.	The very tiny, loose pieces of rock that cover beaches, deserts, etc.1 Made of tiny pieces of rock. 2
Sapling	n.	A young tree.1
Scraps	n.	Fragments of discarded or leftovers.1
Seed Coat	n.	An outer protective covering of a seed.1
Seedling	n.	A young plant that is grown from seed.1 The first sprouts from a seed. They produce the parts of the plant that will be needed for the adult plant to survive in its habitat. 2
Senses	n.	One of the five ways in which you receive information about the world around you, including touch, taste, smell, sight, and hearing.1
Separate	v.	To cause to stop being together, joined, or connected.1
Serving Size	n.	An amount of food or drink that is enough for one person.1
Severe	adj.	Very bad, serious, or unpleasant.1
Shovel	n.	A tool with a long handle that is used for lifting and throwing dirt, sand, snow, etc.1
Signals	n.	An event or act which shows that something exists or that gives information about something.1
Silt	n.	Sand, soil, mud, etc., that is carried by flowing water and that sinks to the bottom of a river, pond, etc.1 Soil that has pieces that are smaller than sand, and therefore feels like powder. 2
Slender	adj.	Small or limited in amount or size.1
Small	adj.	Few in number or little in amount.1
Smooth	adj.	Having a flat, even surface.1
Snowy	adj.	Having a lot of snow.1
Soil	n.	The upper layer of earth that may be dug or plowed and in which plant grow.1 The loose, top layer of Earth's surface made up of pieces of rock, sand, water, air, and pieces of dead organisms, and is used to grow plants. 2
Soil Profile	n.	A vertical section of soil from the ground surface downwards to the underlying rock.1
Solution	n.	Something that is used or done to deal with and end a problem.1 A type of mixture in which one substance is dissolved in another, such as sugar water. 2

Sprouted	v.	To produce new leaves, buds, etc.1
Standard Units of Measure	n.	A unit of measure that is officially accepted as a means to express a quantity.1
Stem	n.	A primary axis of a plant that develops buds and shoots instead of roots.1 The part of the plant that grows out of the ground, supports the leaves, flowers, and fruit, and carries water from the roots to the rest of the plant. 2
Stoma ta	n.	Any of various small simple bodily openings especially in a lower animal.1
Structures	n.	The way that something is built, arranged, or organized.1
Substance	n.	A material of a particular kind.1
Sunlight	n.	The light of the sun.1 Needed by plants to make their own food and grow.
Sunny	adj.	Having plenty of bright sunlight.1
Survival	n.	The state or fact of continuing to live or exist especially in spite of difficult conditions.1
Survive	v.	To continue to function or exist.1
System	n.	A group of related parts that move or work together.1 A group of two or more organs that work together to perform a specific function. 2
Tablespoon	n.	A spoon that is used by cooks for measuring dry and liquid ingredients and that holds an amount equal to ½ fluid ounce or three teaspoons.1
Teaspoon	n.	A spoon that is used by cooks for measuring dry and liquid ingredients and that holds an amount equal to 1/6 fluid ounce or 1/3 tablespoon.1
Technology	n.	The use of science in industry, engineering, etc., to invent useful things or to solve problems.1
Tepee	n.	A tent that is shaped like a cone and that was used in the past by some Native Americans used as a house.1
Terrestrial	adj.	Living or growing on land instead of in water or air.1
Texture	n.	The way that something feels when you touch it.1 Describes the way something feels to the touch. 2
Thunderstorm	n.	A storm with lightning and thunder.1 A severe storm with lightning, thunder, heavy rain, strong winds, and perhaps hail. 2

Toad	n.	A small animal that looks like a frog but has dry skin and lives on land.1
Topsoil	n.	The upper layer of soil in which plants have most of their roots.1
Transpiration	n.	The passage of watery vapor from a living plant through a membrane.1 The loss of water through the leaves of a plant. 2
Transport	v.	To carry something from one place to another.1
Tree	n.	A usually tall plant that has a thick, wooden stem and many large branches.1
Tree Cookie	n.	Cross sections of tree trunks that foresters manufacture, and teachers use to illustrate how trees grow.1
Trowel	n.	A small tool with a curved blade that is used by a gardener for digging holes.1
Underground	adv.	Below the surface of the earth.1
Upward	adv.	From a lower place or level to a higher place or level.1
Variety	n.	A number or collection of different things or people.1
Vegetable	n.	A usually herbaceous plant, such as a cabbage, bean, or potato, that is edible.1
Veins	n.	Any of the tubes forming part of the blood circulation system of the body, carrying in most cases oxygen-depleted blood toward the heart.1
Vermiculture	n.	The cultivation of annelid worms especially for use as bait or in composting.1
Vitamins	n.	A natural substance that is usually found in foods and that helps your body to be healthy.1
Waning	v.	To appear to become thinner or less full.1
Water	n.	The clear liquid that has no color, taste, or smell, that falls from the clouds as rain that forms streams, lakes, and seas, and that is used for drinking, washing, etc.1 One of our most valuable resources on Earth, which every living thing needs to survive. It covers most of the Earth, but only a small portion of it can be used for drinking. It is a liquid that takes the shape of its container. 2
Watering Can	n.	A container usually with a spout used to sprinkle or pour water especially on plants.1
Waxing	v.	To increase in phase or intensity.1

Weed	n.	A plant that tends to overgrow or choke out more desirable plants.1
Whole Grains	n.	Whole grains or foods made from them contain all the essential parts and naturally-occurring nutrients of the entire grain seed in their original proportions.1
Wind	n.	A natural movement of air of any velocity.1
Wind Gust	n	Gust of wind.1
Windy	adj.	Marked by strong wind or by more wind than usual.1
Wood	adj.	The hard substance that makes up the stems and branches of trees and shrubs.1
Wool	n.	The soft, thick hair of sheep and some other animals.1
Worm	n.	A long, thin animal that has a soft body with no legs or bones and that often lives in the ground.1
Worm Castings/ Vermicast	n.	A biologically active mass containing thousands of bacteria, enzymes, and remnants of plant materials and animal manures that were not digested by the earthworm.1
Worm Compost/ Vermicompost	n.	A dark mixture of worm castings, organic material, and bedding in varying stages of decomposition, plus the living earthworms, cocoons, and other organisms present.1

NOTES

CPSIA information can be obtained
at www.ICGtesting.com
Printed in the USA
JSHW041926220223
38112JS00002B/4

9 781949 979978